Wallnig/Evered · Englisch für Baufachleute / L'anglais dans le bâtiment

Günter Wallnig Harry Evered

Englisch für Baufachleute
L'anglais dans le bâtiment

Band 1 • Band 2 • Band 3

Ausgabe von Band 1
für finnische und niederländische Leser

Englantia rakentajille
Engels voor Bouwkundigen

BAUVERLAG GMBH · WIESBADEN UND BERLIN

Günter Wallnig Harry Evered

Englisch für Baufachleute
L'anglais dans le bâtiment

Band 2

3., durchgesehene Auflage

BAUVERLAG GMBH · WIESBADEN UND BERLIN

CIP-Kurztitelaufnahme der Deutschen Bibliothek

Wallnig, Günter:
Englisch für Baufachleute = L'anglais dans le bâtiment / Günter Wallnig ; Harry Evered. – Wiesbaden ; Berlin : Bauverlag
Früher außerdem im Verl. Ed. Eyrolles, Paris

NE: Evered, Harry :

Bd. 2. – 3., durchges. Aufl. – 1986.

1. Auflage 1976
2. Auflage 1977
3. Auflage 1986

ISBN 978-3-528-01703-3 ISBN 978-3-663-05784-0 (eBook)
DOI 10.1007/978-3-663-05784-0

VORWORT

Wegen der großen Nachfrage des 1. Bandes und auf vielfachen Wunsch unserer Leser haben wir diesen Ergänzungsband geschrieben. Wir sind uns darüber klar, daß auch er nicht das volle Bauprogramm umfaßt. Doch glauben wir, mit den ausgewählten Kapiteln dem größten Teil unserer auslandsorientierten Leser eine weitere wertvolle Arbeitshilfe an die Hand gegeben zu haben.

PREFACE

Because of the considerable demand for our first volume and the need to cover further sections of the Construction Industry, we have written this second volume. Whilst we realise that we have still not dealt with all the aspects of construction, we sincerely believe that our second volume will also prove useful and instructive to those interested in technical English.

PREFACE

En raison de la grande demande du 1er volume et selon de multiples désirs de nos lecteurs nous avons écrit ce volume complémentaire. Nous sommes parfaitement conscients que celui-ci non plus n'embrasse pas le programme complet de construction. Mais nous pensons avoir donné à la plus grande partie de nos lecteurs orientés vers l'étranger, avec ces chapitres choisis, une aide de travail précieuse.

Wallnig *Evered*

CONTENTS

1. BUILDING PLANT AND EQUIPMENT

Today's fast developing construction industry depends increasingly on effective and reliable plant and equipment, both on the site and in the workshop. To meet these requirements, designers and manufacturers have already gone a long way in producing greatly improved machinery, but there still remains much scope for further improvements. Far too frequently, inclement weather forces some site plant to a standstill, thereby reducing its profitability, and this is certainly one of the foremost problems facing designers, manufacturers and contractors. Plant involves contractors in a great deal of initial capital expenditure, therefore it is essential that the best possible use should be made of it.

The ability of site machines to withstand the strain of modern construction is an essential factor in their design. For that very reason the vast majority are sturdily built and designed for long life with the minimum of maintenance. The study of the forces acting on earth-moving and other large machines, which is known as soil dynamics or soil vehicle mechanics, has produced a more scientific approach to the design of site machinery. Interest has grown in this subject and one can, therefore, expect that it will have a profound influence on future design.

The large building exhibitions, both on the Continent and in Britain, clearly indicate the growing trend towards increased automation, which results in greater expedience and ease in handling machines. The progress in making machines more compact, as well as safer and more comfortable for the operator, is also very evident. Quite sophisticated heating and ventilating systems, to mention but one refinement, are incorporated in many cabins, thus removing much of the hardship of winter working. The wide range in power and performance provided by manufacturers makes it possible for machinery to be employed with advantage on numerous operations in varying conditions. In fact, even on the smaller and comparatively primitive building sites, some mechanisation, in one form or another, has now become part and parcel of the construction process.

It would certainly appear at the first glance that most site machines these days are hydraulically controlled, and that the old widely employed system of cable control for many operations is gradually disappearing. In fact, most of the modern designed average size construction machines, which are usually powered by diesel engines, are hydraulically controlled. For large capacity machines, such as massive earthmovers, which are intended for high performances, electric power is often adopted. Petrol combustion engines still remain in common use as the mode of power for the small site machines, like tampers, concrete mixers and hoists, although in this field, the development of reliable small diesel engines is leading to the progressive replacement of petrol engines.

Vocabulary	Wörterverzeichnis	Glossaire
construction industry	Bauindustrie, Bauwirtschaft	industrie de la construction
plant	Maschinen und Geräte	matériels
equipment	Ausrüstung	équipement
site	Baustelle	chantier
workshop	Werkstatt	atelier
designer	Konstrukteur	constructeur
manufacturer	Hersteller	fabricant
contractor	Unternehmer	entrepreneur
strain	Beanspruchung	sollicitation
sturdy, rugged	stark, kräftig,robust	robuste
maintenance	Unterhaltung, Wartung	entretien
earth-moving machine	Erdbaumaschine	engin de terrassement
expedience	Zweckdienlichkeit	efficacité

performance	Leistung	rendement
to be part and parcel of . . .	ein wesentlicher Bestandteil sein von . . .	être une partie intégrante de . . .
to control	steuern	commander
cable control	Seilsteuerung, Seilbetrieb	commande à câble
construction machine	Baumaschine	machine pour la construction
to power	antreiben	entraîner
electric power	elektrische Energie	énergie électrique
petrol	Benzin	essence
combustion engine	Verbrennungsmotor	moteur à combustion interne
tamper, earth tamper	Stampfer	pilonneuse
concrete mixer	Betonmischer	bétonnière
hoist	Aufzug	monte-matériaux

"All right, Simpkins. We're all impressed with how clever you are. Now let's get back to work!"

(Reproduced by permission. Masters Agency U.S.A.)

(a) Excavators

Excavators, in their many forms, probably represent one of the largest categories of machinery employed on construction sites. Of these, the universal cable operated excavator is, without doubt, an extremly popular type, although it is gradually being superseded by a similar, but more up-to-date, hydraulically operated machine. Its popularity is due to its versatility, in that it is capable of performing several different tasks through its various interchangeable attachments.

The universal excavator can also be utilised as the base machine for a crane, pile driver, earth tamper, drilling rig, demolition unit and several other items of plant. Because of its many uses in all conditions, this particular machine must be skillfully designed and of sound construction. Already its development has reached a high standard and the scope of its performance greatly widened. It is even possible these days to obtain such a machine giving relatively silent working for use in built-up areas, where noise would prove undesirable.

Basically, an excavator consists of an undercarriage, a superstructure and a boom, or jib as it is also known, together with its attachments. The undercarriage travels either on crawler tracks or rubber tyred wheels, but in each case the ground clearance must be sufficient to enable versatile working of the machine. In order to withstand the most arduous site conditions, the chassis, or frame, of the undercarriage must be particularly robust and completely torsion free. The power to drive the undercarriage is provided by the main engine in the superstructure by means of shafts, gears and roller chains, or through separate hydraulic engines. Fig. 1 illustrates a fully hydraulic excavator, having an air-cooled eight cylinder diesel engine, two hydraulic pumps and a total working weight of about 50 tons.

The principal advantages of crawler tracks, or caterpillar tracks as they are also termed, are that they afford considerable stability on all types of ground, they provide comparatively smooth travel on rugged terrain and permit safe turning, even on fairly steep slopes. Tracks having wide pads, also known as grousers, are particularly suitable on soft ground, because they distribute the machine's weight over a larger area and thus reduce the pressure exerted on the ground. Ideally, it should be possible to replace or adjust the tracks quickly to avoid uneconomical waste of valuable machine working time. The idlers, rollers and drive sprockets, or tumblers in some cases, must be designed for lifetime maintenance-free lubrication and protection against the penetration of dirt.

Rubber tyred wheels also have their advantages, in that they permit a greater travelling speed and enable the excavator to be driven on the public highway with only limited restrictions. On heavy duty excavators the wheel axles are constructed in such a way as to make a high carrying capacity possible. A favourable weight distribution on to both front and rear axles provides much of the necessary stability for the machine in all working positions. Four-wheel drive and large twin-wheels are common features on the majority of machines today, and both are advisable. As with cars, the wheels are provided with two independent braking systems, namely the hydraulically operated four-wheel foot brake and the mechanically lockable hand brake.

The superstructure is supported by and freely revolves on a turntable, which incorporates a slewing ring of ball or roller bearings, and further support is often provided by a sturdy pivot column. The superstructure consists of the mechanism, the operator's cabin, in which the controls for driving, steering, slewing and the manipulation of the various attachments are housed, and the detachable counter balance. The all-round clear visibility afforded by a modern cabin is one of the marked advancements in excavator design. Not only does it contribute directly to more efficient working, but it also achieves a greater degree of safety, and this is an aspect of the building industry which always arouses much public interest.

Years ago the effective operation of an excavator largely depended on the skilful handling of the operator. Up to a point this is still true today, but much of the strain has now been eliminated by automation. The operational functions are usually at the finger tips of the operator by means of a few levers, frequently fitted with air valve thumb triggers to ensure smooth control. Because it is essential during particular operations, the brakes are often designed so that they are held in the release position by springs, but can be locked down by engaging a ratchet. Pressure on the top of the pedal locks the brakes, whereas pressure on the bottom releases them.

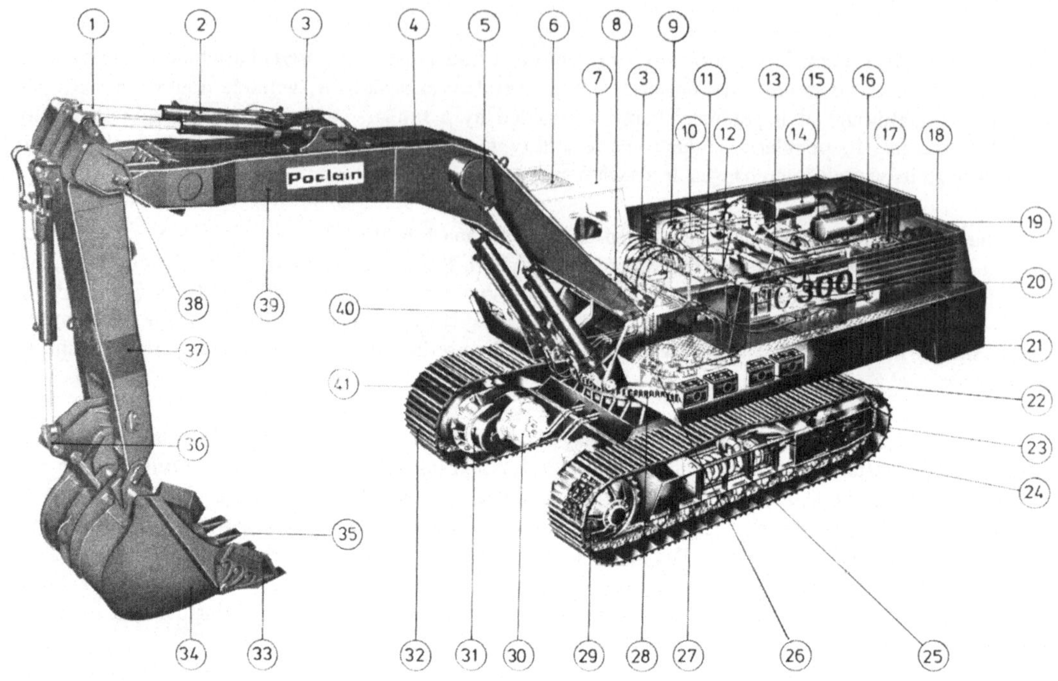

Fig. 1: POCLAIN Hydraulic Excavator

1	**piston rod**	Kolbenstange	tige de piston
2	**ram, cylinder**	Zylinder	cylindre de vérin hydraulique
3	**high pressure hoses**	HD-Schläuche	tuyaux haute pression
4	**pipes**	Rohre	tuyaux
5	**swivel joint**	Dreh(gelenk)verbindung	articulation rotative
6	**protective cover**	Schutzabdeckung	protection
7	**cabin**	Fahrerhaus	cabine du conducteur
8	**control block**	Steuerblock	organe distributeur
9	**cooler**	Kühler	refrigérant
10	**V-belt drive**	Keilriemenantrieb	entrainement par courroie trapézoïdale
11	**pump**	Pumpe	pompe
12	**universal joint**	Kreuzgelenk	cardan
13	**silencer, exhaust muffler**	Auspufftopf	pot d'échappement
14	**air filter**	Luftfilter	filtre à air
15	**diesel engine**	Dieselmotor	moteur diesel
16	**cooler for main engine**	Kühler für Hauptmotor	radiateur du moteur thermique
17	**oil tank, oil reservoir**	Öltank, Ölbehälter	réservoir d'huile
18	**cowling**	Verkleidung	capotage
19	**superstructure**	Oberwagen	tourelle
20	**angular drive**	Winkeltrieb, Umlenker	boîte de renvoi
21	**counter balance, counterweight**	Gegengewicht	contrepoids
22	**battery**	Batterie	batterie
23	**idler**	Leerlaufturas, Spannrad, Umlenkturas	barbotin de renvoi

24	**track tensioning cylinder**	Kettenspann-Zylinder	cylindre de tension de chenille
25	**shock absorber spring**	Stoßdämpfer-Feder	ressort amortisseur
26	**hydraulic slewing motor**	hydraulischer Schwenkmotor	moteur hydraulique de rotation
27	**track supporting roller**	Stützrolle für Raupenkette	galet porteur
28	**slewing gear wheel**	Zahndrehkranz	couronne de rotation
29	**sprocket driving wheel**	(Antriebs-)Turas	barbotin moteur
30	**hydraulic driving motor**	hydraulischer Fahrwerkmotor	moteur hydraulique de translation
31	**planetary reduction gearing**	Planetengetriebe-Untersetzung	réducteur à planétaires
32	**crawler track, caterpillar track**	Raupenkette	chenille
33	**side cutter**	Seitenschneide	lame latérale
34	**backacter, backhoe bucket**	Tieflöffel	godet rétro
35	**replaceable tooth**	auswechselbarer Zahn	dent amovible
36	**universal joint linkage**	Kreuzgelenk-Verbindung	cardan
37	**backacter arm, dipper arm**	Tieflöffel-Pendelarm	balancier
38	**pivot pin**	Gelenkbolzen	axe d'articulation
39	**boom, jib**	Ausleger	flèche
40	**headlight**	Scheinwerfer	projecteur
41	**undercarriage, chassis**	Rahmen, Unterwagen	châssis porteur

The functional equipment of an excavator should always be suitable and readily available for any specific operation, and the interchanging of the various attachments must be simple and rapidly executed. So that the most profitable use can be made of the equipment, the boom should be adjustable to give a variety of working positions. The most utilised attachments, which are normally standard equipment supplied with the basic machine, are the face (or crowd) shovel, backhoe, also termed dragshovel, grab bucket and dragline bucket. Other common attachments are the skim shovel, ditching bucket, hydraulic hammer and lifting hook. Besides these, other most useful and favoured attachments are earth drilling rigs, using augers or drilling buckets, and piling hammers for driving foundation piles.

The most widely employed engine is the powerful diesel combustion type, normally mounted at the rear of the excavator, where it also acts as part of the counter balance. With a mechanically operated machine, all functions are directly executed via the transmission system and gears, whereas with the hydraulic version, these functions are carried out through a hydraulic pump with automatic power control driven by the diesel engine. In the latter case, separate hydraulic working cylinders are provided for digging and loading, as well as hydraulic engines for the travel and slewing necessary throughout these two operations.

The movement of the excavator itself is normally effected through two or four independently operated hydraulic motors, whether it travels on tracks or wheels. Modern designed excavators often provide various combinations of high and low speed gears – e.g. four high speed forward, four high speed reverse, two low speed forward and two low speed reverse. In order to isolate the main engine, whilst it is still running, from the other working machinery during maintenance, it is fitted with a disc clutch. It may sometimes be equipped with a fluid clutch, to cushion it against shocks and to prevent engine stalling. An automatic torque converter is a desirable asset where optimum performance is required, in that it provides smooth operation even at low speeds and under strain. A further development, which has had some useful influence on the design of hydraulic excavators, is the axial piston pump. Such pumps are most effective in producing the necessary high breakout forces, which can be exerted at the bucket teeth when encountering hard material in the excavations.

Where electricity is the mode of power for large machines, it is usually provided by a diesel engine or a three-phase alternating current (a. c.) motor driving, directly on the same axle, a direct current (d. c.) generator, which supplies one or more motors for the various operations. This method, known as the "Ward-Leonard" system, is extremely efficient and incurs low maintenance costs. The revolutions, the active power, and the most effective torque can be easily and widely varied without gears, by means of electrical resistance controllers. For standard electrically powered machines, a three-phase rotary current motor is usually employed.

Other forms of excavating machines, perhaps not so common, but nevertheless worthy of some mention, are bucket excavators and dredgers. In the most modern type of the former, the digging operation is performed by a series of buckets rotating as a wheel at the end of the jib. The excavated material is tipped on to a conveyor belt on an independent jib and dumped into a spoil heap or directly into a haul unit. Both these functions on the larger machines are powered electrically. The excavation of long deep trenches or the cutting of canals are practical propositions with such a machine.

Dredgers, which are floating vessels used principally for clearing mud and other deposits from canal and river beds, normally make use of a chain system of buckets for lifting the dredged material. However, the suction dredger, using large centrifugal pumps and large diameter suction pipes is rapidly superseding this long established method. This has already been extensively borne out, for example, in the Dutch Delta Scheme, in which harbour basins were formed and at the same time the level of the surrounding terrain was raised.

Vocabulary	Wörterverzeichnis	Glossaire
excavator	Bagger	excavateur
versatility	Vielseitigkeit	versalité, maniabilité
attachment	Anbaugerät	équipement interchangeable
crane	Kran	grue
pile driver	Pfahlramme, Ramme	sonnette de battage
drilling rig	Bohrgerät	engin de forage
demolition unit	behandeltes Gebiet	engin de démolition
built-up area	geschlossene Bebauung	aire bâtie
undercarriage	Unterwagen	chassis porteur
superstructure	Oberteil, Oberwagen	tourelle
boom, jib	Ausleger	flèche
crawler track, caterpillar track	Raupenkette	chenille
tyre	Reifen	bandage
wheel	Rad	roue
ground clearance	Bodenfreiheit	garde au sol
shaft	Welle	arbre
gears	Zahnräder, Getriebe	roues dentées
roller chain	Rollenkette	chaine à galets
grouser, track pad	Kettenstollen	crampon
idler	Leerlaufturas, Umlenkturas	barbotin de renvoi
drive sprocket	Antriebsturas	barbotin
tumbler	Schwenkzahnrad, Kettenstern	roue dentée orientable
lifetime lubrication	Dauerschmierung	graissage à vie
maintenance free	unterhaltungsfrei, wartungsfrei	sans entretien
heavy duty	schwere Ausführung	poids lourd
wheel axle	Radachse	essieu
twin-wheel	Zwillings-Reifen	bandages jumelés
braking system	Bremssystem	système de freinage
lockable brake	Feststellbremse	frein de parcage

turntable	Drehscheibe	plaque tournante
slewing ring	Drehkranz	couronne de rotation
ball bearing	Kugellager	roulement à billes
pivot column	Schwenkzapfen	pivot
operator's cabin	Fahrerhaus	cabine du conducteur
control	Steuerorgan, Bedienungsorgan	organe de commande
to drive, to travel	fahren	rouler
slew	schwenken	tourner
to house	unterbringen	mettre à l'abri
detachable	abnehmbar, demontierbar	amovible
counter balance	Gegengewicht	contrepoids
all-round visibility	Rundumsicht	visibilité totale
lever	Hebel	levier
air valve	Druckluftventil	soupape d'air
thumb trigger	Taster, (Finger-)Druckauslöser	bouton-poussoir
release position	Ruhestellung	position de repos
to lock, to lock down	verriegeln	verrouiller
ratchet	Zahnstangensperre, Zahnklinkensperre	rochet
adjustable boom	umrüstbarer Ausleger	flèche amovible
face shovel, crowd shovel	Hochlöffelbagger, Hochlöffel	pelle pour travail en butte
backhoe, dragshovel	Tieflöffelbagger, Tieflöffel	rétrocaveuse
grab bucket	Greifer	benne preneuse
dragline bucket	Schleppschaufel, Schürfeimer	benne à traction
skim shovel, skimmer	Planierlöffel	pelle niveleuse
ditching bucket	Graben (profil)-Löffel	godet de tranchée
lifting hook	Lasthaken	crochet de levage
earth drilling rig	Erdbohrgerät	foreuse-sol
auger	Schneckenbohrer	tarière
drilling bucket	Schürfbohreimer	godet de forage
piling hammer	Rammbär	mouton
foundation pile	Gründungspfahl	pieu de fondation
diesel combustion engine	Diesel-Motor	moteur Diesel
power control	Leistungsregelung	réglage de la puissance
speed gear	Getriebegang	rapport de vitesse
speed reverse	Rückwärts(gang)-Geschwindigkeit	rapport arrière
disc clutch	Scheibenkupplung	embrayage à disque
fluid clutch	hydraulische Kupplung	embrayage hydraulique
to cushion	abpuffern, dämpfen	amortir
to stall	abwürgen	caler
torque converter	Drehmomenten-Wandler	convertisseur de couple
axial piston pump	Axialkolbenpumpe	pompe à piston axiale
three-phase alternating current motor	Dreiphasen-Wechselstrommotor	moteur à courant triphasé
direct current, (d. c.)	Gleichstrom	courant continu
revolutions per minute, r.p.m.	Drehzahl	régime
torque	Drehmoment	couple
resistance controller	Regelwiderstand, Steuerwiderstand	résistance réglable
rotary current	Drehstrom	courant rotatoire
bucket excavator	Schaufelradbagger	excavateur à roues à godet
dredger	Schwimmbagger, Naßbagger	drague
to tip	kippen	basculer
conveyer belt	Förderband	transporteur à courroie
to dump	abkippen	déverser
spoil heap	Kippe	lieu de décharge
to haul	transportieren	transporter
haul unit	Transportfahrzeug	véhicule de transport

trench	Graben	tranchée
suction dredger	Saugbagger	drague suceuse
harbour basin	Hafenbecken	bassin portuaire

Dialogue

Duncan: Tell me, Robert, which item of excavating equipment do you find most widely used on the Continent?

Robert: It's difficult to say exactly, but if you are thinking of the large site works where vast quantities of soil are being moved, I would think it's the dragline.

Duncan: Of course, it requires much skill on the part of the operator and it can prove very economical in good hands. If you think it would interest our readers, I'll try to outline what else I consider to be essential in the effective operation of the dragline.

Robert: Go ahead, Duncan, but just before you begin, I must say I can fully appreciate the snags in controlling the swinging bucket dangling by cables alone from the light lattice box boom.

Duncan: It's not at all easy, as you have already pointed out. Without the correct type of bucket, well maintained, it is virtually impossible to achieve good results. It must have the proper shape and weight for the particular soil, it's cutting teeth should be kept really sharp and it is essential that the bucket is held in exactly the correct position by the cables.

Robert: I'm glad you mentioned the cables, for this is an opportune moment to say something about the vital part that they play. All three cables, that is the hoist line, drag line and dump line, the first two of which are wound up on the winch drums, have to be controlled simultaneously. Incidentally, the dump line is sometimes referred to as the tip line. In the normal operating sequence, the bucket is lowered, rather than thrown, into the pit, with the hoist and drag lines taut. Just as the bucket is about to strike the ground, the hoist line brake is almost completely released, and then re-applied gradually at the same time as the drag line brake is released. In this way, the bucket drops vertically. The hoist line is now slackened a little and the drag line clutch engaded, whereby the bucket is pulled along by the drag line and so cuts into the ground, removing a slice in the process. The final operation of dumping the excavated soil certainly requires as much skill as the excavation itself.

Duncan: With some training, of course, a competent operator would be capable of carrying out the whole process almost instinctively.

Robert: Of all the special devices used on modern designed excavators, we can hardly afford to overlook the one employed in the excavation of narrow and generally deep trenches to accommodate diaphragm walling.

Duncan: Doubtless, you are referring to the slot-cutting clamshell grab with a particularly long guide-bars assembly for excavating widths of about 60 centimetres. Both cable and hydraulic control are possible, as you know, but as a matter of interest, which method do you consider is preferable?

Robert: It all depends upon the maximum depth of dig required, and in this respect, a cable suspended clamshell grab has the advantage of permitting far deeper excavation. It should be remembered, however, that a good deal of protection against corrosion from the Bentonite mud is essential for the bearings of the clamshell pivot joint, as well as for the rope sheaves and guide rollers. This means, in fact, that they must be effectively sealed against the likely penetration of the mud.

Duncan: This protective measure is equally important for those vulnerable moving parts of the hydraulically controlled clamshell grab (or dragshovel in some situations) and the greatly extended dipper arm. Incidentally, the possible depth of dig attainable with such an arm can vary from about 10 to 14 meters below ground level.

Robert: Before moving on to another topic of interest, are there any other particularly important attachments about which you would like to say a few words?

Duncan: Actually, there are quite a number, Robert, but I'll confine myself to just a couple, which I feel warrant a little discussion. The lifting of large boulders, which is often beyond the capability of even the largest and sturdiest bucket, necessitates the use of rather special attachments. This difficulty is normally overcome by means of a grapple bucket or a crampoon, both of which are hydraulically controlled.

Robert: These devices are especially useful, of course, in quarry excavation, where large blocks of stone require their jaw-like grip to shift them. The double acting rams of the crampoon are most effective for handling outsize blocks. Switching the subject somewhat, Duncan, do you believe that the future holds anything for cable controlled excavators?

Duncan: Well, I certainly don't think they are destined to an early extinction, but I feel we are bound to see fewer of them in years to come. Their simplicity and straight forward maintenance are well proven, and with this in mind, some contractors will always favour them.

Robert: For that reason, I understand, the Americans still manufacture them to satisfy their continuing demand.

Duncan: Whatever the future of cable control may be, hydraulics will, I'm sure, eventually sweep the field with the smaller machines, because hydraulic engines have been proved to be reliable. There had been difficulties for some time in the development of a suitable engine for this purpose. The smaller machines must be very economically designed, otherwise many contractors could never afford to employ them.

Robert: Whilst on the subject of future development, would you care to forecast whether most machines of the next decade or so will be mounted on crawler tracks or tyred wheels?

Duncan: As both have their obvious attributes, it's not easy to determine what the possible outcome might be, but I feel that tracks could find most favour, mainly because they provide greater stability.

Robert: Maybe you are right, Duncan, but with overdimensioned tyres, a wheeled mounting is certainly more stable than is normally the case. Furthermore, retractable hydraulic stabilizer feet, or outriggers, used on wheeled machines undoubtedly eliminate instability in stationary working positions.

Duncan: They are, in fact, beams which are drawn out from the chassis to form supports for the undercarriage. Used in conjunction with jacks or blocks, they provide a rigid working base and permit a greater lifting capacity than even a crawler excavator of comparable size.

Robert: An interesting development on one particular German wheel-mounted hydraulic excavator is its ability to operate on railway track, as well as on normal sites. It is equipped with four railway wheels on two axles, which are hydraulically retractable.

Duncan: This is certainly a very useful machine for the railway, both for laying ballast around the sleepers and for clearing obstructions from the track, as a result of landslides etc. Another double-purpose hydraulic excavator, which springs to mind, is an Italian model. In this case, both the wheel-mounted and crawler-mounted versions can be easily converted into a loader, by simply removing the excavator arm and replacing it by a rugged loading shovel provided with twin arms.

Robert: Reverting back to the movement of excavators, one usually thinks only of tracks or wheels, but perhaps we shouldn't overlook the walking frame, even though it is seldom used with excavators. It moves rather like a baby on all fours, as you may know. A baby progresses by moving its arms forward and resting its weight on the hands whilst swinging its body forward with the feet dragging behind. Then it rests on the knees and toes before moving the arms forward again.

Duncan: The principle, of course, could have a greater use in future large excavators, where the absolute stability of the machine is of utmost importance.

Robert: I realise I'm wandering somewhat from the point, but it's worth mentioning a special ram of German origin, which embodies this particular stepwork. The undercarriage consists of a wide shallow cylinder, which rests on the ground and carries the superstructure. The cylinder is lifted clear of the ground by four shoes acting like jacks, which protrude from the walking frame and enables the machine to move in any direction to a new working position.

Duncan: As you are well aware, the majority of foundation, drainage and cable trenches etc. are dug out with a hydraulically operated backhoe, but the trend today, both on the Continent and in Britain, is moving towards the excavation of such trenches from the side of the excavator. In other words, this is leading to a fuller use of off-centre attachments.

Robert: That's quite true, Duncan. Positioning the digging bucket to one side of the revolving superstructure permits mechanical excavation close to existing buildings or other obstructions. This is possible either with a cranked jib or one which moves, to a certain degree, from one side of the machine to the other by means of a joint and hydraulic cylinders.

Duncan: An extremely useful attachment, which in time should eliminate yet another awkward excavation job, hitherto only possible by hand. Incidentally, what is your impression of the typical trencher with a digging wheel, around the perimeter of which are attached a series of buckets or shovels? Personally, I haven't come across one for some time.

Robert: Well, frankly, I think this type of machine is only suitable for specific purposes, such as drainage ditches and bulk excavation work.

Duncan: In mentioning the latter purpose, I take it you are referring to the huge bucket wheel excavators employed in open cast coal mining?

Robert: Yes, I am in fact. Furthermore, I would mention, that many years ago I actually witnessed three of these giants in operation in such a coal mining area between Cologne (Köln) and Aix-la-Chapelle (Aachen). Each one soared some 70 meters into the sky, had an enormous weight approximately equivalent to that of an old steel bridge crossing the Rhine, and through its 120 electro motors, consumed as much electricity as a town having about 50,000 inhabitants.

Duncan: Excavators constitute a rather wide subject, which we could undoubtedly discuss until the cows come home. I think your last piece of descriptive information will serve admirably to round off our discussion, so may I suggest, Robert, that we break off for the moment.

Vocabulary	Wörterverzeichnis	Glossaire
snag	Schwierigkeit, Haken	difficulté
lattice box boom	Gitterausleger	flèche en treillis
hoist line	Hubseil	câble de levage
drag line	Zugseil	câble de cavage
dump line, tip line	Kippseil	câble de renversement
winch drum	Windentrommel	tambour de treuil
operating sequence	Betriebsfolge, Arbeitsspiel, Takt	phases de fonctionnement
pit	(Bau) Grube	fouille
taut	straff, angespannt	tendu
to slacken	locker lassen, lockern	relâcher
to engage a clutch	einkuppeln	embrayer
to release a clutch	auskuppeln	débrayer
diaphragm walling	Schlitzwand	mur emboué
slot-cutting	Schlitzaushub	excavation pour mur emboué
clamshell grab	Schalengreifer	benne à demi-coquilles
guide-bars assembly	Führungsgestänge	barres de guidage
depth of dig	Grabtiefe	profondeur de cavage
pivot joint	Drehzapfengelenk	articulation à pivot
rope sheave	Seilrolle	poulie à gorge
guide roller	Führungsrolle	galet de guidage
to seal	abdichten	étancher
dipper arm	Löffelstiel, Auslegerarm	bras de godet de pelle
boulder	Steinbrocken	pierre lourde
grapple bucket	Mehrschalengreifer, Polypgreifer	benne à mâchoires multiples
crampoon	Zangengreifer	griffe de levage
quarry	Steinbruch	carrière
jaw	Backenklaue	mâchoire
grip	Klemmkraft	force de serrage
to shift	verschieben	déplacer
double-acting	zweiseitig arbeitend, doppelt beaufschlagt	à double action
contractor	Unternehmer	entrepreneur

to sweep the field	den Markt beherrschen	dominer le marché
attribute	charakteristische Eigenschaft	caractéristique
wheeled mounting	Radunterwagen	chariot
outrigger	Stützarm	vérin de calage
jack	Hebebock, Wagenheber	vérin
block	Klotz	bloc
wheel-mounted excavator	Radbagger	excavatrice à roues
railway track	Eisenbahngleis	voie ferrée
ballast	Schotter	ballast
sleeper	Schwelle	traverse
obstruction	Hindernis	obstruction
landslide	Erdrutsch	éboulement de masse
loader	Lader	chouleur
stepwork	Schreitwerk	mécanisme promeneur
perimeter	Umkreis, Umfang	périmètre
drainage ditch	Entwässerungsgraben	fossé de drainage
joint	Gelenk	articulation
bulk excavtion	Massenaushub	excavation en masse

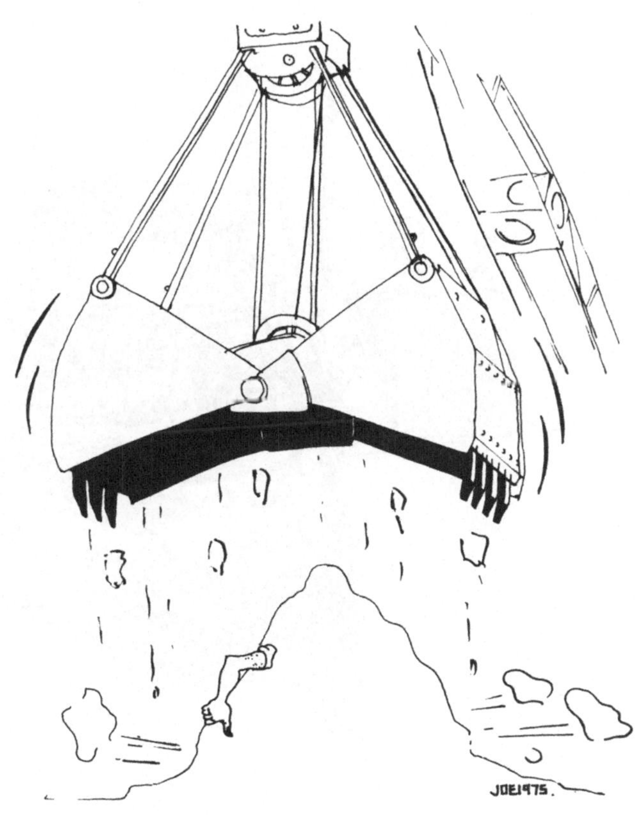

"Over th . . . .

(b) Dozers and Loaders

Besides excavators, there are several other machines widely engaged in earthmoving and operations of a like nature, and of these, dozers and loaders (known also as tractor shovels) are doubtless the most useful for multi-purpose work. Dozers are employed mainly to push away earth and other obstacles, in order to reduce ground levels, whilst loaders are generally used to excavate and remove various materials from spoil heaps and deposit them elsewhere. The popularity of these machines is well established, and this is repeatedly borne out by the fact that one or the other invariably becomes a contractor's first investment in site mechanisation.

Both the dozer and loader have been developed from the ordinary tractor and, like excavators, are designed to run either on crawler tracks or wheels, according to the site conditions and their actual duties. Ideally, the basic machine for the dozer should be different from that for the loader, because the forces acting on the chassis are dissimilar in each case. However, for economic reasons and to reduce the variety of spare parts, manufacturers usually mass produce a number of basic machines, serving as a compromise between the specific requirements of each type. The fundamental difference between these two earthmovers and the excavator is that in the former the working attachments are directly connected to the machine itself, whereas in the latter they form part of the slewing superstructure.

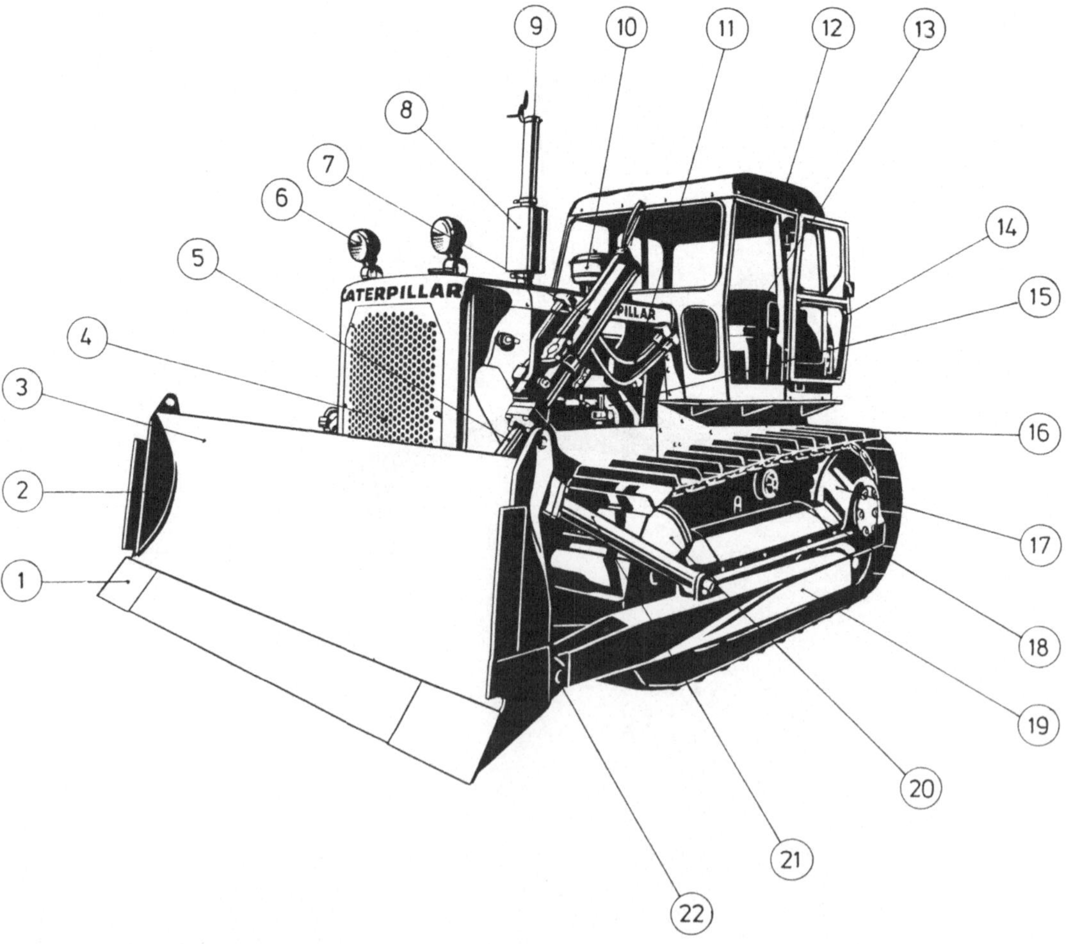

Fig. 2: CATERPILLAR Bulldozer

bulldozer	Flachbagger mit Querschild	bulldozer

1	**replaceable cutting edge**	austauschbare Schneidkante	trousse coupante interchangeable
2	**segmental blade end**	Schildseitensegment	segment de lame
3	**bulldozer blade**	Schild, Querschild	lame
4	**radiator grill**	Kühlerschutzgitter	grille de radiateur
5	**piston rod**	Kolbenstange	tige de piston
6	**headlight**	Frontscheinwerfer	phare de route
7	**double-acting ram**	doppeltwirkender Hydraulikzylinder	vérin double effect
8	**silencer**	Auspufftopf	pot d'échappement
9	**exhaust pipe outlet**	Auspuffrohrende	extrémité du tuyau d'échappement
10	**air cleaner**	Luftfilter	filtre à air
11	**flexible high pressure hose**	Hochdruckschlauch	tuyau flexible à haute pression
12	**cabin hood**	Fahrerhausdach	toit de cabine
13	**operator's sprung seat**	gefederter Fahrersitz	siège à ressorts du conducteur
14	**hinged cabin door**	Fahrerhaus-Klapptür	porte articulée de cabine
15	**4-stroke diesel engine with 4 cylinders**	4-Zylinder-4-Takt-Dieselmotor	moteur Diesel 4 cylindres à quatre temps
16	**crawler chain, caterpillar track**	Raupenkette	chenille
17	**driving sprocket**	Antriebsturas	barbotin moteur
18	**supporting roller**	Stützrolle	galet de support
19	**thrust frame**	Schubrahmen	châssis pousseur
20	**idler**	Leerlaufturas, Umlenkturas	barbotin de renvoi
21	**blade tilting arm**	Schildkippstrebe	bras de basculage de la lame
22	**lower blade hinge bolt**	unterer Schildanschlußbolzen	boulon de connexion inférieur lame

The entire working cycle of many of the older machines still in use today necessitates skilful handling, for which the operator requires nerves of steel, as well as complete physical fitness and mental alertness. He must keep the machine in motion when working hard, without stalling the engine, because of its low kinetic energy compared, for example, with a moving car, and must also control the working operations, such as dozing, digging, loading, lifting and dumping. The modified design of most modern dozers and loaders has greatly simplified most operations, so much so, in fact, that the average car driver, given some special training, could probably handle one with a reasonable degree of efficiency and safety.

Within this group of earthmovers, the bulldozer is the oldest and simplest form. As it moves forward, on tracks in most cases, the blade cuts into the ground, pushes the excavated material along and diverts it to each side for subsequent disposal. The blade on this particular machine extends straight across the front beyond the tracks or wheels, is slightly concave in section and often has narrow segmental ends. It is normally pivoted on two sturdy arms connected to the thrust frame of the undercarriage and manipulated by double-acting hydraulic lift and tilt cylinders, either directly or through intermediate links. A bulldozer is often utilised to push or pull scrapers to assist their loading, and when pushing, it is normally fitted with a sprung blade to cushion the impact. Other duties performed by bulldozers from time to time include towing various rollers and dump trailers. Although cable controlled blades, using winches and wire cables, do have their merits, the hydraulically actuated types have far-reaching advantages, and consequently the former are seldom seen these days.

When moving soft earth, the bulldozer blade is tilted slightly backwards from the vertical position, whereas it is tilted forward when tackling harder material. Either corner of the blade can usually be dropped lower than the other, in order to move stubborn boulders which hamper progress. When the going is really heavy, necessitating the lowest gear, the blade should be gradually raised to reduce the friction and avoid a considerable slowing down of the engine or undue spinning of the tracks or wheels. Here, much operating skill is called for and it is best carried out instinctively and not abruptly, which would only result in an uneven excavated strip. The operation of moving material on a slope should, wherever possible, be performed downhill, thus making use of the natural force of gravity.

The sturdy bulldozer blade, which can be fitted on practically all dozers, is ideal for most short distance earthmoving operations in the toughest conditions. However, where it is necessary to move more material over greater distances, an angledozer is more suitable than a bulldozer. With an angledozer, the straight blade is simply replaced by a longer blade, extending far beyond the tracks or wheels, having open sides and set at an angle to the thrust frame. Pushed soil offers less resistance to an angled blade than to a straight one, and thus it is possible to provide the former with a greater length. As a result of this, wider strips can be excavated at a greater speed. The maximum angle usually found to be most effective is about 30°. Because the long angled blade, which is fixed well out in front, tends to move up and down with the movement of the machine, precise excavation on undulating ground is not possible. Since the excavated material is piled up on one side of the machine only, its later disposal is made easier. It is possible to angle-doze to a lesser extent with the majority of ordinary bulldozer blades, because they can be quickly repositioned on the thrust frame to serve as a short version of the angledozer blade.

The loader, being a rehandling machine, is principally employed to transfer material, excavated by some other machine, from the point of deposit to a waiting haul unit a short distance away. In selecting a loader, it is important to take into full account its stability and mobility, both of which are critical factors. Its maximum payload, utmost dump height and maximum reach are three of the main features of considerable interest to a contractor, especially in terms of profitability. Because of the versatility of the loader, it is invariably overworked, and it must therefore be very robustly constructed. The resistance against any tendency for a wheeled loader to roll backwards when handling material depends largely on the correct weight distribution over the wheels, the size and tread of the tyres and the total weight of the machine. In specific instances, some manufacturers even advise contractors to fill the inner tubes of the tyres with water rather than compressed air, so as to increase the overall weight and stability. Some wheeled machines are provided with two-axle oscillation to ensure positive down pressure on all wheels at all times.

Basically, there are three types of loader, namely the front, overhead and side loader, which differ one from the other in the method of dumping the payload. The front loader is by far the most common, being quite adequate in most situations. The advantage of the rear or side loader is most prominent in confined spaces, where the movement of the machine itself is somewhat restricted. In the case of the rear loader, the bucket, or shovel as it is also termed, travels over the loader and is emptied at the rear. With the side loader, the bucket can also be tipped to one side to deposit its contents. This particular machine needs to be rather more heavily and sturdily constructed, in order to withstand the additional torsion acting on the various parts.

The loading bucket, up to a point, resembles the bulldozer blade, the main difference being that it has totally enclosed sides. A wide range of buckets, both in size and type, are produced to meet the various requirements, and many have removable teeth or cutting edges. Most loaders are equipped with at least four hydraulic cylinders, two for lifting and lowering and two for tilting the bucket. On some modern models, the operation of the bucket arms and bucket is controlled through one lever, rather like the joy stick of an aeroplane. The loading capacity, lifting force and maximum thrust must be properly matched, and these three factors depend, above all, on the performance of the engine, known in Britain as the net flywheel horsepower. In the U.S.A., such performance is normally given an S.A.E. rating, S.A.E. referring to the Society of Automobile Engineers.

In addition to the blade, which constitutes the basic equipment of a dozer, other useful attachments are often employed. Amongst these, some of the most common are rippers, winches, rock rakes, cable or pipe plows, and slopers. With regard to the loader, on which the bucket represents the main attachment, the secondary attachment at the rear is nearly always a slewing back-acter complete with a trenching bucket. Many of these back-acters are designed to move along a carriage beam for the full width of the loader, which enables them to excavate from offset positions. However, rippers, tampers, breaking hammers, pile drivers, and post hole augers, amongst others, may sometimes act as alternative pieces of additional equipment. To provide stability for a wheeled loader when a back-acter is in operation, the front bucket is usually forced against the ground, so lifting the front wheels, whilst the rear wheels are raised by hydraulic jacks, which are designed to suit uneven as well as level ground.

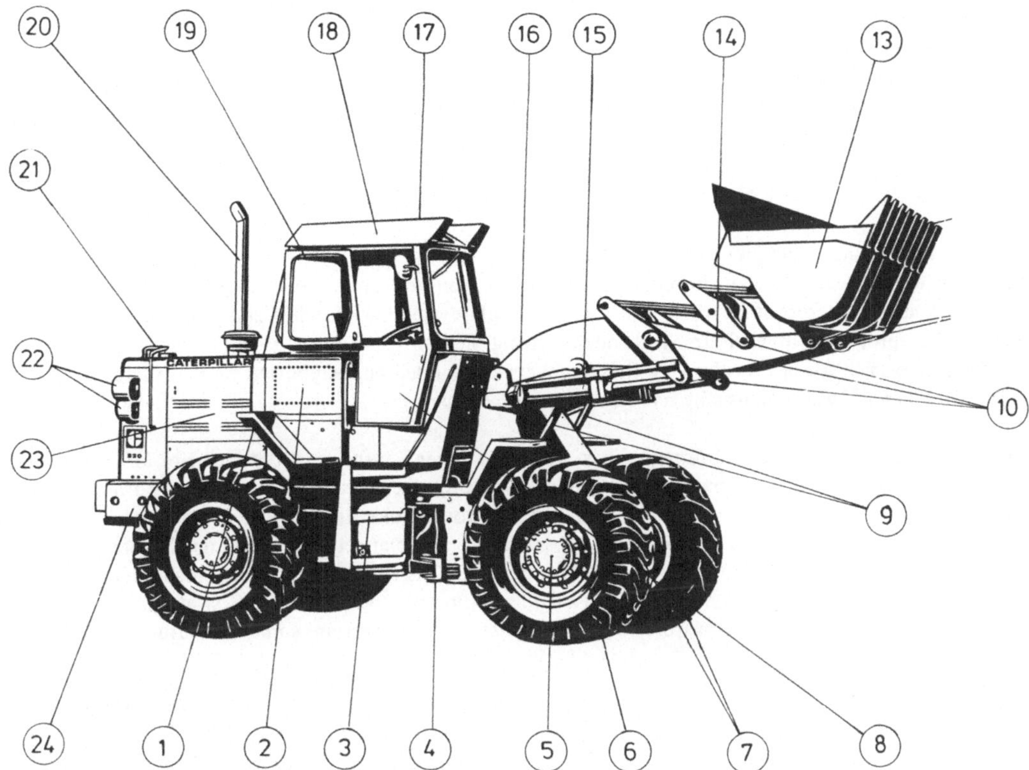

Fig. 3: CATERPILLAR Wheel Loader

	wheel loader, tractor shovel	Radlader	chouleur-pelleteur à roues, chargeuse sur roues
1	**mudguard, wing**	Kotflügel, Schutzblech	garde-boue
2	**hydraulic oil tank with oil level gauge**	Hydrauliköltank mit Ölstandsichtglas	réservoir d'huile de pression avec jauge de niveau
3	**step of access ladder**	Leiterstufe	marche d'échelle
4	**lower pivot bearing of articulated steering**	unteres Zapfenlager der Knicklenkung	pivot inférieur bas de la direction pivotante
5	**wheel hub with planetary reduction gear**	Radnabe mit Planeten-untersetzungsgetriebe	moyeu avec réduction planétaire
6	**wheel rim**	Radfelge	jante
7	**tyres of the front wheels**	Vorderradreifen	pneus des roues avant
8	**cabin door**	Fahrerhaustür	porte de cabine
9	**piston rods of lift cylinders**	Kolbenstangen der Hubzylinder	tiges de piston des vérins de levage
10	**tilting linkage bearings**	Kippgestängelager	appuis de la tringlerie de basculement
11	**bucket joint bearings**	Schaufelgelenklager	roulements de l'articulation de la benne
12	**bucket tooth**	Schaufelzahn	dent de benne
13	**loader bucket, face shovel**	Ladeschaufel, Frontladeschaufel	benne
14	**bucket lifting arm**	Schaufelhubarm	bras de levage de la benne
15	**hydraulic tilting ram**	(hydraulischer) Kippzylinder	vérin de basculement
16	**headlight**	Frontscheinwerfer	phare avant
17	**operator's cab**	Fahrerhaus	cabine
18	**working light**	Arbeitslampe	phare de travail
19	**open cabin side window**	geöffnetes Fahrerhausseiten-fenster	fenêtre de cabine ouverte

20	**exhaust pipe**	Auspuffrohr	tuyau d'échappement
21	**radiator filler cap**	Kühlerverschluß	bouchon du radiateur
22	**stop and tail lights**	Brems- und Rückleuchten	feu de freinage et feu rouge
23	**bonnet, engine cowling**	Motorhaube, Motorverkleidung	capot
24	**chassis, undercarriage**	Chassis	châssis

The essential driving components of the basic machine for the older types of common crawler dozer or loader are a comparatively low powered diesel engine, mounted on a welded mild steel chassis, a main single-plate clutch, a robust gearbox and two multi-plate steering clutches and steering brakes, which operate the sprockets for each track through driving shafts. The tracks, as with crawler-mounted (or tracklaying) excavators, are revolved by the sprockets and guided and supported by idlers and rollers. The brakes, clutches and gears are usually hand-operated through levers.

The call for better working performances has, to a large extent, influenced the inclusion of a number of notable improvements into the more recent models. One prominent example is the more powerful modern engine, which in most cases is an air- or water-cooled four-stroke diesel type, designed to run with a continuous rating exceeding 100 horse power (h. p.) at about a relatively low 2,000 revolutions per minute (r.p.m.) and having a comparatively low fuel consumption. This engine requires less maintenance, than its predecessor, has fewer replaceable wearing parts and seldom fails to start at the first attempt, even at low temperatures. As with some excavators, the larger dozers and loaders are sometimes driven by hydraulic engines.

The transmission of power from the engine to the tracks has improved remarkably. The former friction clutch, which often proved to be somewhat fierce in operation, is now often replaced by a turbocoupling, – in other words a hydrodynamic drive, which permits smooth starting, devoid of stalling. Alternatively, a torque converter enables the operator to adapt and alter both the r.p.m. and the torque. The combination of a torque converter and a robust gear transmission, or planetary gearing, provides a full powershift range of speed by optional torque. Compared with the older transmission systems, the modern one eliminates much of the unnecessary trouble formerly experienced by operators, since all operations are hydraulically actuated. Because he requires only one lever to select two or three ranges of speed, or reverse, the operator always has one hand and both feet completely free to control the steering and working equipment, which in themselves present no operational difficulties nowadays.

A further marked improvement appears in the chassis, which these days is normally constructed of high tensile creep-resistant steel box-section framework. The tracks too have greatly improved, in that most now have lifetime maintenance-free self-lubricating bearings and can easily be removed or interchanged for tracks of different widths. Recent design permits the track assemblies of larger machines to be speedily removed, in order to reduce the overall width to conform to the regulations for transportation on a standard low-loader. The smooth travel of the tracks, even over normal obstacles, is considerably enhanced by their being fully oscillating, and with the provision of self-adjusters and heavy duty shock absorbers incorporated with the rollers and idlers. The oil compartments of the tracks are hermetically sealed against the leakage of oil and ingress of dirt, and both the preservation and noise abatement of the tracks is assisted by rubber protected rollers.

Generally speaking, the basic principles of the driving system for wheel-mounted machines are similar to those described for crawler mounted types. In the former, however, the drive torque is transmitted to the drive shafts via spur and bevel gears. For example, by using a three-speed full reversing constant-mesh gear with hydraulic clutches, all shifts, up or down, in forward or reverse, can be made on the move. No stopping for range shifts is necessary. On most of these machines, all four wheels are driven by power-transfer differentials. It is normally possible by the movement of a lever to disconnect the drive of one axle when the machine is merely travelling along the road, or wherever four-wheel drive is not justified. The differential is normally locked automatically as soon as one of the wheels begins to slip, and this ensures that each wheel constantly transfers its full share of engine torque to the ground. Quite a number of modern machines are provided with planetary reduction in each wheel, which provides more torque at the wheels, where all available traction and power are required.

The steering of a crawler-mounted machine is normally carried out by using one of the clutching and braking systems provided for each track. A single movement of a hand-lever firstly releases the clutch and then follows this up by applying the brake. For example, to turn to the right, the right hand lever is applied, and at the same time the left hand track sprocket is driven. In such a manoeuvre, a torque converter is of great assistance, in that it reduces the need for extreme care in handling. On smaller machines, the driving sprockets are often driven through a differential gear, as with a car, and this makes turning simple. In this case, a right hand turn is carried out by merely applying the right hand brake through a lever or wheel.

An important aspect of the wheel-mounted machine is its manoeuvrability, because this, above all, is a time-saving factor in the working cycle. Some steering systems are derived from that of a car, but usually act on the rear wheels and are always hydraulically assisted. A very simple, yet efficient, form of steering linkage consists of twin hydraulic rams acting on the wheels via a bellcrank and drag links. A further step towards improved manoeuvrability is in the form of the articulated power steering system, which is fast becoming increasingly popular. In this case, the rams act directly on the separated front or rear part of the machine. This part is linked to the main part by a heavy pivot pin, which enables the axle of the smaller part to turn almost 45°, giving an extremely small turning circle. The contra-rotation of wheels, as with tracks too, is easily possible with modern machines, and this provides added mobility in confined spaces.

Illustrations of an average sized "Caterpillar" crawler bulldozer and wheel loader are given in Figs. 2 and 3 respectively. The bulldozer is designed to utilise a wide variety of attachments and is easy to maintain. Other features are its high ground clearance, whether with direct drive or power shift transmission, and its sealed tracks. The versatile loader has a number of important characteristics, namely power shift transmission controlled by a single lever, torque converter, all-wheel drive with planetary reduction on each wheel and hydraulic power articulated steering.

The comfort of the operator should never be overlooked, because fatigue through discomfort can seriously hamper the safety and productivity of the machine. In this respect, manufacturers have endeavoured in recent times to provide many improved facilities to remove former irritations and laboriousness from the routine operating procedure. Cushioned bucket seats, adjustable to suit operators of varying stature, is a notable feature in many modern cabs, or cockpits as they are also known. The careful positioning of pedals and levers, to be within easy reach and to give an uncluttered floor area for speedy access and freer movement, has proved to be of some benefit, and clearly defined dial panels help to eliminate any operational tensions. The panoramic view afforded by the large sliding or pivoted windows of up-to-date cabs lends itself to greater safety and more relaxed working. On a number of machines, the cab is constructed of plastic within stout metal framing or of fibreglass and can easily be removed in fine weather to give a completely unobstructed view. Few cabs today are without heating and air conditioning systems, or have not been insulated against much of the outside noise. In fact, the whole design of a cab nowadays, contributes to faster and more efficient working cycles.

Progress to date in the development of dozers and loaders has been quite significant, and in view of the fact that there are now so many manufacturers of these machines throughout the world, an even faster rate of progress can be expected.

Vocabulary	Wörterverzeichnis	Glossaire
dozer	Flachbagger	bulldozer
loader, tractor shovel	Lader, Ladeschaufler	chouleur-pelleteur, chargeuse
working cycle	Arbeitsspiel	cycle de travail
dozing	Bodenschürftransport	bulldozage
bulldozer	Flachbagger mit Querschild (Brustschild)	bulldozer

blade	Schild	lame
disposal	Abfuhr	évacuation
pivoted	angelenkt	articulé
thrust frame	Schubrahmen	châssis pousseur
to tilt	kippen	basculer
intermediate link	Zwischenglied	membre intermédiaire
scraper	Radschrapper, Schürfkübelwagen	décapeuse à roues
sprung blade	gefederter Schild	lame suspendue à ressorts
to tow	ziehen	traîner
roller	Walze	rouleau
dump trailer	Anhängekipper, Nachlaufkipper	remorque basculante
cable control	Seilsteuerung, Seilbetrieb	commande par câbles
winch	Winde	treuil
wire cable	Stahlseil	cable d'acier
gear	Gang	rapport
to spin	durchdrehen	patiner
angledozer	Seitenräumer	bulldozer à lame orientable
undulating ground	welliges Gelände	terrain onduleux
payload	Nutzlast	charge utile
dump height	Schütthöhe	hauteur de vidage
tread	Lauffläche	bande de roulement
inner tube	Luftschlauch	chambre à air
overall weight, total weight	Gesamtgewicht	poids total
front loader	Frontlader	chargeur frontal
overhead loader, rear loader	Überkopflader	rétrochargeur
side loader	Seitenlader	chargeur latéral
cutting edge	Schneidkante, Schneide	tranchant de lame
joy stick	Steuerknüppel	levier de commande
to match	(an)passen, (ab)stimmen	adapter
performance	Leistung	rendement
net flywheel horsepower	Nettoleistung in PS	puissance nette au volant
ripper	Aufreißer	scarificateur
b.h.p., brake horsepower	Brems-PS (etwa DIN-PS)	puissance au frein
rock rake	Steinrechen	râteau à pierres
cable or pipe plow	Kabel- oder Rohrverlegepflug	dispositif de pose des canalisation ou de câbles
sloper	Böschungshobel	lame de talutage
back-acter	Tieflöffelbagger	rétrocaveuse
trenching bucket	Grabenlöffel	godet rétro-tranchées
carriage beam	Gerätebalken, Schlittenträger	poutre porteuse
tamper	Stampfgerät	pilonneuse
breaking hammer	Abbruchhammer	marteau de démolition
pile driver	Pfahlramme	sonnette de battage
post hole auger	Pfostenlochbohrer	foreuse pour pose de poteaux
wheeled loader	Radlader	chargeuse à roues
hydraulic jack	hydraul. Stempel	vérin
crawler dozer	Raupenfronträumer, Raupenflachbagger	bulldozer chenillé
to weld	schweißen	souder
mild steel	Flußstahl	acier doux
single-plate clutch	Einscheiben-Kupplung	embrayage monodisque
gearbox	Getriebe	mécanisme
multi-plate clutch	Mehrscheiben-Kupplung	embrayage à disques
sprocket	Turas	barbotin
track, crawler chain	Raupenkette	chenille

driving shaft	Antriebswelle	arbre moteur
idler	Leerlaufturas, Umlenkturas	barbotin de renvoi
roller	Rolle	galet
gear	Getriebegang	rapport
4-stroke engine	4-Taktmotor	moteur à quatre temps
fuel consumption	Kraftstoff-Verbrauch	consommation de carburant
wearing part	Verschleißteil	piece d'usure
transmission of power	Kraftübertragung	transmission
friction clutch	Reibungskupplung	embrayage à friction
drive	Antrieb	entraînement
to stall	abwürgen	caler
gear transmission	Zahnrad-Getriebe	engrenage
planetary gearing	Planetengetriebe	train planétaire
powershift gear (gear)	Lastschaltgetriebe	boîte de vitesses automatique powershift
range of speed	Drehzahlbereich	régime
high tensile	hochzugfest	à haute résistance
creep-resistant	kriechfest	résistant au fluage
box-section	Kastenprofil	profilé en forme de caisson
to lubricate	schmieren	graisser
bearing	Lager	roulement
overall width	Gesamtbreite	largeur hors-tout
regulation for transportation	Transport-Bestimmung	réglementation des transports
low-loader	Tieflader	remorque surbaissée
to oscillate	pendeln	osciller
self-adjuster	automatischer Nachsteller	dispositif autoréglable
shock absorber	Stoßdämpfer	amortisseur de chocs
heavy duty	schwere Ausführung	poids lourd
leakage	Lecken	fuite
noise abatement	Lärmminderung	protection antibruit
spur gear	Stirn(zahn)rad	roue dentée droite
bevel gear	Kegelrad	roue conique
full reversing	voll umkehrbar	complètement réversible
constant-mesh gear	Dauereingriff Getriebe	engrenage en prise constante
range shift	Gangschalten	changement de la vitesse
power transfer differential	Kraftübertragungsdifferential	différentiel
differential lock	Differentialsperre	blocage de différentiel
planetary reduction	Planetenuntersetzung	réduction planétaire
hydraulically assisted steering	hydraulische Lenkhilfe	direction à assistence hydraulique
steering linkage	Lenkgestänge	tringlerie de direction
hydraulic ram	Hydraulikzylinder	vérin
bellcrank	Winkelhebel	levier coudé
drag link	Lenker, Lenk(schub)stange	barre de direction
articulated steering	Knicklenkung	direction articulée
power steering	Servo-Lenkung	servo-direction
to link, to articulate, to hinge	gelenkig verbinden	articuler
pivot pin, pivot column	Gelenkzapfen	pivot
attachment	Anbaugerät	équipement interchangeable
to maintain	unterhalten	entretenir
ground clearance	Bodenfreiheit	garde au sol
drive	Antrieb	dispositif entraîneur
power shift transmission	Lastschalt-Getriebe	boîte de vitesses automatique powershift
sealed track	(lager)gekapselte Laufkette	chenille étanchéisée

torque converter	Drehmomentwandler	convertisseur de couple
all-wheel drive	Allradantrieb	entraînement toutes roues
cushioned	(gegen Stöße) abgefedert, gepolstert	amorti
bucket seat	Schalensitz	siège baquet
cab, cockpit	Fahrerhaus	cabine
dial panel, instrument panel	Instrumentenbrett, Armaturenbrett	tableau de bord
sliding window	Schiebefenster	fenêtre coulissante
pivoted window	Schwingflügel-Fenster	fenêtre battante
fibreglass	Faserglas, glasfaserverstärkter Kunststoff	fibre de verre
air conditioning system	Klimaanlage	système de climatisation

Dialogue

Robert: I'm sure you share my view, Duncan, that within this particular section, we are all now witnessing what must surely be the fastest development of site machinery in the building industry.

Duncan: Indeed, I do. In fact, I would go as far as to say that some of the items discussed today could well be out of date in a short space of time. Maybe we should, therefore, concentrate some of our discussion on future trends in dozers and loaders.

Robert: Agreed, but before doing so, may I say just a little about one or two of the important features, which I believe a prospective buyer would most look for in a modern machine.

Duncan: Yes, by all means, Robert, go ahead, for it should be of some interest to most readers.

Robert: Naturally, one would expect maximum output to be foremost in the contractor's thoughts, but mind you, his attention would also be focused very much on both the initial and running costs of the machine.

Duncan: To my way of thinking, Robert, you've hit on three crucial factors. There's little need to say any more about costs – they speak for themselves, but would you please enlarge somewhat on output.

Robert: Well, in the case of a loader this is determined, to a certain extent, by its safe payload, which is usually represented by the struck capacity of the bucket, whereas with the dozer, it is partly dependent on the blade thrust. Of course, other factors come into this, as you know.

Duncan: May I please interrupt you at this convenient point to briefly mention a couple of these, which I consider to be of some importance. I'm thinking principally of fast cycle times, even over the worst terrain, and constant machine working, with the minimum of downtime for repairs and maintenance, which contribute considerably to output. It may appear all too obvious, but I believe it's well worth stressing, that however efficient a machine may be, much of the benefit gained whilst its working at full capacity can so easily be whittled away when it stands idle for long periods.

Robert: You've just mentioned repairs and maintenance, Duncan. Well, both should only be necessary now and again and their execution must be rapid to reduce wasted time, which a cost-conscious contractor would regard as eating into the profits.

Duncan: Before turning our attention to another topic, we shouldn't forget the greatest expectation of all purchasers, that their machine will give many years of trouble-free service together with effortless operation in all conditions. This may sound like an idealistic dream, but surely it could be possible with all the advanced techniques of today.

Robert: I've no doubt you're right, and in my opinion many modern machines do, in fact, already fulfil the greater part of this desire. I know that there are obstacles which stand in the way of progress towards the ultimately perfect machine – that is, as perfect as possible.

Duncan: You intrigue me, Robert, with your remarks, so please continue.

Robert: Unfortunately, economics influence manufacturers, just as they do contractors, and in consequence all the technical know-how available to a producer is not always utilised, because of the cost. One example I can give you of economic aspects taking precedence over technical ones, is in the development of the fully hydraulic layout of a machine, which has decided advantages over the mechanical system, yet at the moment is not an economic proposition. I am convinced that other brilliant ideas like this are thwarted and that the ideal machine does exist in the mind of the designer, on the drawing board, or even possibly at the prototype stage. Getting such a machine on to the market presents umpteen problems, as you can imagine, the foremost being recovering the development expenses. In competition with the mass produced models of the giant concerns, known as 0815-machines in Germany, I believe, it wouldn't stand a dog's chance of success.

Duncan: What you have said is very true, even if it sounds a little pessimistic, but research and development is rarely wasted in the long run and outstanding ideas do eventually filter through to appear even in the mass-produced machines. Take for example the single lever control of all bucket movements or the non-stop power shift transmission, both of which inevitably spark off interest with customers. It stands to reason that features like these either become universally adopted or they lead to even better ideas to attract custom. Through this process, the perfect machine may one day emerge.

Robert: Reverting back to power shifting, which you touched on a moment ago, I would point out that this is only possible because of the application of a hydraulic torque converter. I have, however, heard of certain drawbacks in this system.

Duncan: Ah, yes! you are thinking, no doubt, of the losses in energy, which, I admit, are normally greater with a hydraulic transmission system. Despite this, you can't escape the fact that progress in this field has been quite remarkable.

Robert: I certainly wouldn't dispute that, Duncan, and I'll leave it to you to provide some supporting data.

Duncan: Very well, here goes! Take for example the efficiency of an up-to-date torque converter or modulation coupling. This is as high as 94 to 96 per cent.

Robert: That sounds good, but perhaps we should really explain at this point what exactly a modulation coupling is. It is, in fact, a multi-disc clutch activated by a large section piston, combined with a torque converter and pump. The unit provides an ideally adaptable transmission. In other words, it ensures that the operator can't possibly make gearing mistakes, because the unit actually thinks for him. The most satisfactory result of all this is that the operator can devote more attention to the digging or loading job.

Duncan: Switching to something rather different, what is your re-action, Robert, to the possible danger created by the ever increasing pressures in the hydraulic systems of new models? I feel, personally, that the operator would run a considerable risk of serious injury, if the system ever became damaged.

Robert: To be quite honest, I believe there is no need for concern on this matter, provided that the best quality high pressure hoses and pipes are used. I would think that this hazard has been pretty well mastered. In fact, on some models banks of neatly arranged steel pipes with safety couplings can be seen.

Duncan: Before going on any further, we must bring up the subject of rippers, for these are in such common use both on caterpillar and wheel-mounted machines.

Robert: I entirely agree, Duncan. They are extremely useful attachments for breaking out hard material, prior to the main excavation or loading operation.

Duncan: It's quite incredible how they can sometimes handle even huge chunks of solid rock. It shouldn't be forgotten, of course, that a great deal of their capability depends on the drawbar pull of the tractor.

Robert: Granted, some credit should be given to the tractor, but the rippers themselves have reached a high standard of proficiency.

Duncan: As you are doubtless aware, there are quite a variety of modern rippers, but broadly speaking they fall into two basic types, namely the radial lift and parallel lift. They are termed as such from the mechanics employed to raise and lower the tool beam. Some rippers which are mounted in a fixed position still exist today, but most have been mainly replaced by the radial or parallel lift type, which are actuated by hydraulic cylinders to force them into the ground and to lift them clear.

Robert: Another variation is in the number of shanks employed. A single shank is quite common, but probably more common is a row of four or five. For scarifying, of course, an even greater number are often used, but each shank, or tooth as it is sometimes called, is somewhat smaller. Vibrating shanks are becoming more popular than their rigid counterpart, because they produce a greater fragmentation of the material being broken up, reduce the shock transmitted to the machine and the operator and inflict less wear on the shank points.

Duncan: Whilst thinking of rippers, I am reminded of the plow, which is, after all, a form of ripper with the added facilities for laying cables or small diameter pipes.

Robert: You've certainly picked out one of the outstanding labour saving attachments. It's not only fitted with lifting cylinders for the special shank, but also has cylinders to adjust its pitch and to move it to either side of the centre line position.

Duncan: There really seems no end to the range of useful attachments and we could chat indefinitely on the subject. However, if you have any last minute thoughts on any other special attachments, maybe you would like to express them.

Robert: Yes, there are a few that justify some comment, in my view. The tree cutter is one. This merely consists of a toothed blade, which is fixed in front of the dozer blade and cuts through the tree as the machine applies pressure. Another is the side boom pipe-laying crane, usually employed only for handling large diameter pipes. Lastly, the multi-purpose 4 in 1 bucket warrants special mention. This attachment, normally used as an alternative to the common loader bucket, can be utilised as a bottom dump or conventional dump loading bucket, a dozer bucket, a scraper bucket or a clam bucket.

Duncan: Well, Robert, I think we've dwelt long enough on attachments and I would suggest we round off our discussion with some interesting points relating to specially built dozers.

Robert: Which ones have you in mind, Duncan?

Duncan: Actually, I was thinking of swamp and underwater dozers. Dealing with the first, it is worth noting the principal differences between this machine and the ordinary dozer.

Robert: The first thing that strikes me is the increased width of the track shoes, or pads as they are also called. This is obviously necessary to reduce the machine's specific ground pressure, considering the extremely soft ground on which it must work.

Duncan: The other noticeable differences, of course, are the greater distance between the tracks and the higher ground clearance. Now to move on to underwater dozers, which are really only in their infancy. The Japanese have already produced a prototype with a diesel engine, which is capable of dozing and ripping 15 metres under water. They have also built an electro motor type, which can work at an even greater depth – up to 60 metres, I understand.

Robert: Yes, I've heard of these two remarkable machines, both of which are either cable or radio controlled. One could say, in fact, that they are robots, rather than machines.

Duncan: Having discussed quite a variety of rather specific items, it would be appropriate, I feel, to close this episode on a more general note. Actually, I'm thinking of the human angle regarding the effect machines have on the community. Even though some steps have already been taken to cut down the noise and fumes, which cause some annoyance to the general public, as well as site personnel, the perfect servant in all respects has yet to be produced. This is certainly one of the important challenges for the future.

Vocabulary	Wörterverzeichnis	Glossaire
output	Arbeitsleistung, Fördermenge	rendement
safe payload	zulässige Nutzlast	charge utile admissible
struck capacity	gestrichener Inhalt	capacité à ras
cycle time	Arbeitsspielzeit	temps de cycle de travail
downtime	(Arbeits-)Ausfallzeit	temps mort

drawing board	Zeichenbrett	planche à dessin
single lever control	Einhebel-Steuerung	commande à levier unique
large section piston	Kolben mit großem Querschnitt	piston à large section transversale
gearing mistake	Schaltfehler	erreur de changement de vitesse
pressure hose	Preßluftschlauch	tuyau flexible pour air comprimé
ripper	Aufreißer	scarificateur
chunk	dicker Brocken	grosse pierre
drawbar, tow bar	Deichsel, Zugstange	barre de traction
drawbar pull	Zugkraft an der Deichsel	effort de traction disponible à la barre
shank	(Reiß-)Schenkel	sabot
to scarify	flach aufreißen	désagréger
scarifying tooth	Aufreißzahn	dent de scarificateur défonçage
fragmentation	Zertrümmerung	fragmentation
wear	Abnutzung, Verschleiß	usure
plow	Pflug	charrue
pitch	Neigungswinkel	pente
bottom dumping	Bodenentleerung	vidage par le fond
dump loading bucket	Ladekipp-Schaufel	godet basculant
clam bucket	Schalengreifer	benne à demi-coquilles
cable control	Fernsteuerung über Kabel	télécommande par câble
radio control	Funksteuerung	télécommande par radio

"I'll take two!"
(Reproduced by permission. ATECO U.S.A.)

(c) Pumping Equipment

There are two main reasons why pumping equipment is so necessary on sites, the first and most usual of which is to keep the open excavations free from excess water, so that reasonably dry working conditions can be maintained throughout the construction of the sub-structure. Waterlogged conditions nearly always result from the soil itself having a high water table, which requires lowering, but they can also be caused by the inability of the soil to drain away heavy or prolonged rainfall by natural means. The second reason is to provide a water supply for general purposes from a temporary tank or natural source, in those situations where a mains supply is not available. Such equipment also has an important function in the pumping of suitable concrete or cement grouting and the lifting of sewage from one point to another. In the case of the latter operation, unchokeable pumps are usually employed.

Basically, the equipment comprises a pump, prime mover, suction hose (unnecessary in the case of a submersible pump), which is fitted with a strainer to exclude large solids, and delivery hose, also termed a discharge hose. The greater part of the pump itself is usually of cast iron or aluminium, and the hoses are normally of flexible wire-armoured canvas-backed rubber or of reinforced plastics. There are quite a variety of pumps, differing in their fundamental working principle and also in their purpose and performance, but the three types most commonly employed in the construction industry are self-priming centrifugal, lift and force diaphragm (single or double diaphragm) and submersible pumps. Other types, less frequently used are semi-rotary, plunger and air lift pumps. The transportation of pumping equipment also varies, many of the heavier types being wheel-mounted, usually with a tow bar, whilst the lighter ones are either skid-mounted or provided with a carrying handle.

The expression "suction head", in relation to centrifugal and diaphragm pumps, refers to the height from the water level to the pump position, and ideally, this should never exceed about 7 meters. "Delivery head" refers to the height from the pump position to the point of delivery. A pump is said to be "on snore" when it is sucking in more air than water, – not at full capacity, in fact, – and the improved design of most centrifugal and diaphragm pumps permits them to run efficiently and without harmful effects or attention for long periods under such conditions. Modern electrically driven submersible pumps are usually fitted with an automatic cut-out device, which stops the pump when there is insufficient water to justify continued pumping.

A centrifugal pump, broadly speaking, is composed of a spiral body, known as a volute, within which an impeller, normally of brass or bronze, rotates at high speed on the driving shaft. The shaft is driven directly by a prime mover, being either an electric motor, petrol engine, diesel engine or even a compressed air motor. Water is forced through the suction hose to the centre of the impeller, from which it is thrown to the periphery by the curved or radial impeller vanes and on into the delivery hose. This type of pump is usually preferred for pumping large quantities of clear water, but it is capable of handling sand and small gravel in water.

A diaphragm pump, which is the most popular form of displacement pump, consists basically of a body, within which a large circular plate having a flexible diaphragm type die-moulded rubber or neoprene flap valve, is forced up and down by a rod. This rod is actuated by the prime mover, which, in this case, can be either an electric motor, petrol engine, diesel engine or, in odd instances, a manually operated lever bar. Water is lifted into the body through the suction pipe and forced out through the delivery pipe by the downward thrust of the plate. The principal advantage of this pump is its ability to handle gritty water, sludge, small stones and even a high percentage of heavy solids in suspension with little effect on the working parts. The larger forms of this pump, having two or three plates, are known respectively as duplex or triplex lift and force diaphragm pumps.

There are two quite distinct, yet common, types of submersible pump, one in which only the actual pump unit is totally submerged in the water, whilst the prime mover, being similar to that used for a centrifugal pump, is out of the water and is connected to the pump unit by means of a length of flexible drive shaft. The smaller prime movers are often mounted on a swivel base, to allow free movement of the drive shaft. The second and more recent type of submersible pump consists of a compact automatic self-contained unit, which houses both the pump and electric motor and which is completely submerged.

In the partially submerged type, the submersible pump unit comprises a pump body, within which the vaned impeller rotates, a hinged strainer fitting over the impeller, and a pump head, which houses the impeller shaft. The action of the revolving impeller forces water through the strainer, drives it to the side of the body and up the delivery hose. The flexible drive is normally connected to the power drive with a quick release plug-in coupling. The pump unit together with the drive shaft is often termed the "pump pac". Corrosion resistant materials are used in the manufacture of submersible units, being usually aluminium alloy for the body, head and strainer, chrome alloyed steel or ductile cast iron for the impeller, stainless steel for the impeller shaft and steel covered with a wire braid reinforced rubber for the flexible drive shaft.

The self-contained submersible pump is a very simple, yet reliable unit. Its compactness and complete absence of protruding working parts makes it ideal for use in confined spaces and on dewatering systems. Such units are invariably provided with a built-in thermo-switch, which cuts out the pump in the event of overloading. Water is forced by the impeller through grille holes in the bottom of the casing and, as with the other type, it is driven to the side and up the delivery hose. To attain a greater delivery head, these units can be connected in tandem. In a number of models both low and high pressure impellers are provided, these being interchangeable. The former are capable of pumping large volumes of water at low heads, whilst the latter can handle smaller volumes at greater heads. Similar corrosive resistant materials are used in a self-contained submersible pump as are used in the other type, with the exception that the unit body is normally made of galvanised steel, which is often corrugated for added strength. The electric current is supplied through a toughened rubber or plastic double-insulated feed cable, having a watertight connection into the unit.

Modern submersible pumps are able to handle sludge and quite heavy solids in suspension, as well as being capable of giving a fairly high output. In fact, some of the larger ones provide the major advantages of both the conventional centrifugal and diaphragm pump.

Fig. 4 shows a detailed cross section of Flygt submersible pumps. Both these electrically driven rotary dewatering units are able to run continuously for weeks without maintenance, apart from a monthly oil level check. There is no risk of damage to these units when running "on snore".

On sites which are seriously hampered by water, such as those immediately adjacent to the sea or inland watercourse, those situated in swampy terrain or those on which the natural drainage is particulary poor, some special form of dewatering system is necessary to ensure tolerably dry working conditions. The principle of such a system is to draw the water downwards through the ground (that is to lower the water table), pump it away, and so leave the ground firm enough to permit normal working without the danger of the sides of the excavation caving in.

In the case of relatively small pit excavations in clay, marl or fissured rock, open dewatering is sometimes possible. This is one of the simplest dewatering systems, and merely consists of a number of sumps excavated around the perimeter of the pit and extending below the bottom of the main excavation. A submersible pump is located in each sump, and thus the water is drawn away from the pit. For larger shallow excavations in clay, fine sand or clayey/sandy soil, the wellpoint dewatering system is often the most suitable, but in medium or coarse sand, a shallow filter well system is preferable.

In the wellpoint system, each individual wellpoint forming part of the general layout, comprises a vertical perforated steel core, usually a metre or so long, enclosed by a brass or other non-corrosive screen, a self-jetting shoe at the lower end and a steel head at the top. Attached to each head is a vertical riser, which in turn is connected to the top of a common horizontal header pipe, usually by means of swings. The header pipe, which encloses an area a little larger than that to be excavated, is directly connected to one or more suction pumps, according to the extent of the dewatering operation and depending on the size of the pumps. The discharging of the water at a conventient position some distance from the site, is carried out as in any normal pumping job. It is possible to control the volume of water drawn from the ground as required, by means of gate valves on the header pipe and cocks on the risers.

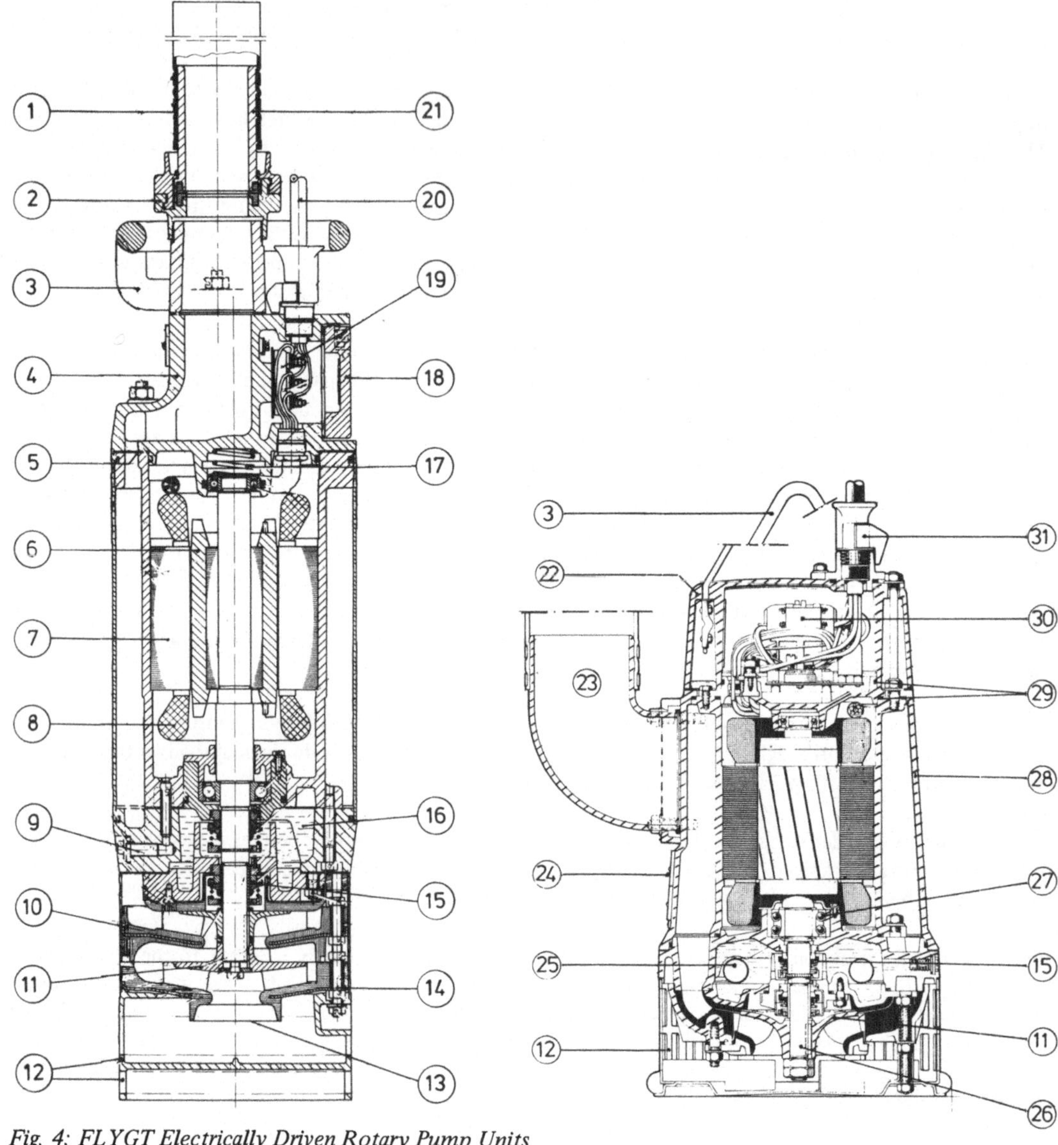

Fig. 4: FLYGT Electrically Driven Rotary Pump Units

1	**hose with hose clamp**	Schlauch mit Schlauchschelle	tuyeau avec collier de serrage
2	**quick-release coupling**	Schnellkupplung	raccord rapide
3	**carrying handle**	Tragbügel	anse
4	**pump head**	Pumpenkopf	tête de pompe
5	**packing**	Packung	joint
6	**squirrel-cage rotor**	Käfiganker, Kurzschlußläufer	rotor à cage
7	**stator of 3-phase induction motor**	Ständer des Drehstrommotors	stator du moteur triphasé
8	**stator windings**	Ständerwicklung	bobinage
9	**inspection screw**	Inspektionsschraube	vis d'inspection
10	**diffuser ring**	Diffusor-Ring	anneau diffuseur
11	**impeller**	Laufrad	rotor
12	**strainer**	Sieb	crépine
13	**inlet opening**	Einlauföffnung	orifice d'entrée

14	**stud with nuts and spring washer**	Stiftschraube mit Muttern und Federring	goujon avec écrous et rondelle de blocage
15	**lower seal unit**	untere Dichtungseinheit	joint mécanique, NF
16	**oil filling**	Ölfüllung	remplissage d'huile
17	**spring**	Schraubenfeder	ressort hélicoïdal
18	**junction box cover**	Anschlußdosendeckel	couvercle de la boîte de raccordement
19	**terminal board**	Klemmenplatte	plaque à bornes
20	**4-core cable**	vieradriges Kabel	câble à quatre brins
21	**pressure pipe**	Druckrohr	tuyau de pression
22	**upper part of the casing**	oberes Gehäuseteil	partie supérieure du carter
23	**outlet bend**	Auslaßkrümmer	coude décharge
24	**data plate**	Typenschild	plaque signalétique
25	**pressure compensator**	Druckausgleicher	compensateur de pression
26	**pump shaft**	Pumpenwelle	arbre de pompe
27	**ball bearing**	Kugellager	roulement à billes
28	**outer casing**	äußeres Gehäuse	enveloppe
29	**O-rings**	O-Ringe	anneaux toriques
30	**contactor unit**	Schützeinheit	contacteur complet
31	**cable entry unit**	Kabeleinführungseinheit	entrée de câble complet

A most efficient operating unit for a wellpoint system, the "Flygt Vakumat", is illustrated in Fig. 5. An interesting and useful feature of this particular unit is the special inlet flange with a suction lock, which permits the pumps to be disconnected or reconnected without loosing vacuum in the system. Another advantage is its economical operation, mainly due to the stepped switching system, which switches individual pumps on or off according to the volume of water and air being handled.

The initial operation in the wellpoint system is to sink the wellpoints to the required depth, this being limited to approximately 7 metres in any one sinking. To reach greater depths, it is necessary to set up a further system, once the first stage of dry excavation has been completed. The actual sinking of a wellpoint is carried out by driving water at high pressure down the riser by means of a high pressure jetting pump. The water is forced through the wellpoint via the ball valve in the shoe, and is emitted in jets from the pointed shoe, thus pushing away soil particles in its path. Under its own weight, plus that of the riser, the wellpoint sinks naturally into the ground. This process is generally known as jetting.

Once the penetration of the wellpoints has progressed sufficiently, the major function of dewatering can commence. Having disconnected the jetting pump and hose, all the components of the system are connected and the wellpoint is put under suction. The ground water is drawn freely through the filtering screen up the centre core of the wellpoint and on through the network of risers and the header pipe, to be eventually discharged from the suction pump.

In the shallow filter well system, each well sunk into the ground consists of an inner suction pipe and an outer filter pipe, and the piping network is similar to that of a wellpoint system.

For deeper excavations in most soils, other than clay, the deep filter well system is usually adopted. Here, each well in the ground consists of an outer filter pipe and an inner steel pressure pipe, at the foot of which is attached a special slender submersible pump. The pump itself comprises three compartments, the lower one being for a three-phase alternating current (a.c.) motor, the middle one for the filter and the uppermost one for a series of horizontal coupled centrifugal pumps.

One of the chief advantages of the wellpoint system and both filter well systems is that they often obviate the necessity for expensive planking and strutting or sheet piling, and this in itself can expedite the excavations. Whatever method of dewatering is employed, it is always necessary to provide a standby pump for use in case of pump failure.

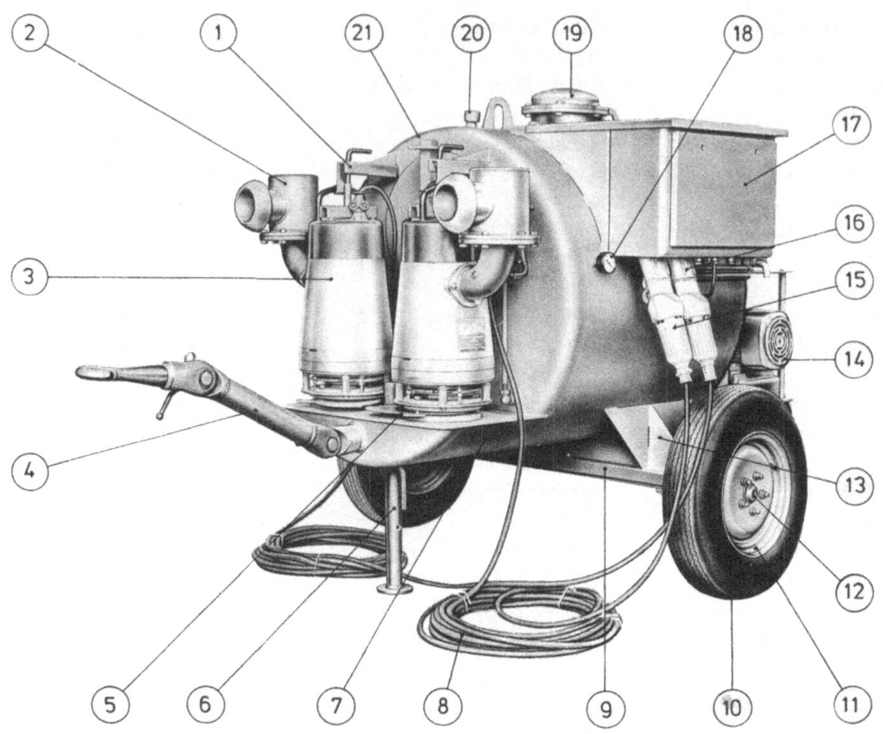

Fig. 5: FLYGT Automatic Electric-Vacuum Unit

1	**clamping device for positioning one pump unit**	Spannvorrichtung zur Montage einer Pumpeinheit	dispositif de serrage pour montage d'un groupe
2	**water outlet pot**	Wasserauslaßtopf	déchargeur
3	**electrically driven rotary pump unit**	elektrisch angetriebene Kreiselpumpeneinheit	groupe centrifuge de pompage électrique
4	**tow bar with joints**	Zugstange mit Gelenken	barre d'attelage articulée
5	**inlet flange piece**	Einlaß-Flanschstück	bride d'entrée
6	**standing leg**	Absetzstütze	support
7	**water filled part of tank**	mit Wasser gefüllter Teil des Tanks	partie de bac remplie d'eau
8	**4-core heavy duty rubber sheathed electric cable**	elektrisches Kabel, starke Gummischlauchleitung, vieradrig	câble à gaine à quatre conducteurs
9	**wheel axle**	Radachse	essieu
10	**tyre**	Reifen	bandage
11	**rim**	Felge	jante
12	**hub**	Nabe	moyeu
13	**stiffening angle**	Versteifungswinkel	profilé raidisseur
14	**driving motor**	Antriebsmotor	moteur entraîneur
15	**3-phase coupling plug**	Kupplungsstecker für Kraftstrom	fiche de prise de courant force motrice
16	**socket coupler**	Kupplungssteckdose	prise de courant
17	**switch box**	Schaltkasten	boîte de manœuvre
18	**vacuum gauge**	Unterdruckmesser	vacuométre
19	**magnetically operated float switch**	Magnetschwimmhalter	interrupteur a flotteur magnétique
20	**safety float valve**	Sicherheits-Schwimmerventil	interrupteur à flotteur
21	**vacuum tank**	Vakuumkessel	bac à vide

Pumps should always be well maintained and their replaceable components easily accessible, so that the risk of failure is small and any unavoidable delay for minor repairs is reduced to the minimum. Breakdowns with such crucial plant for long periods can seriously impede the progress of the work and result in unnecessary expense.

Vocabulary	Wörterverzeichnis	Glossaire
sub-structure	Unterbau, Bauteil unter Gelände	partie au-dessous du terrain
waterlogged	wassergesättigt	saturé d'eau
water table, groundwater level	Wasserspiegel, Grundwasserstand	nappe phréatique
to drain	entwässern, ableiten	drainer
temporary tank	Hilfsbehälter, Baustellentank	bac auxiliaire
mains	Hauptleitung	canalisation
mains supply	Netzversorgung	alimentation par réseau
cement grout	Zementschlämme	coulis de ciment
sewage	Abwasser	eaux usées
unchokeable	verstopfungsfrei	incolmatable
prime mover	Antriebsmaschine, Motor	mécanisme entraîneur
suction hose	Saugschlauch	tuyau flexible d'aspiration
submersible pump; submerged pump	Tauchpumpe	pompe immergée
strainer	Saugkorb	crépine
solid	Feststoff	matière solide
delivery hose, discharge hose	Ablaßschlauch	tuyau flexible de refoulement
cast iron	Gußeisen	fonte
wire-armoured	drahtverstärkt	renforcé par fils métalliques
canvas-backed	segeltuchverstärkt	renforcé par toile de lin
reinforced plastics	verstärkter Kunststoff	plastique renforcé
self-priming	selbstansaugend	auto-amorçant
centrifugal pump	Kreiselpumpe	pompe centrifuge
lift and force diaphragm pump	Saug- und Druck-Membranpumpe, S- und D-Diapumpe	pompe à membrane aspirante et foulante
semi-rotary pump	Flügelpumpe	pompe à ailettes
air lift pump	Mammutpumpe	pompe mammouth
wheel-mounted	auf Rädern, auf einem Fahrgestell	sur roues
tow bar	Deichsel, Zugstange	barre d'attelage
skid-mounted	auf Kufen	sur châssis kid
carrying handle	Traggriff	anse
suction head	Saughöhe	hauteur d'aspiration
delivery head	Druckhöhe	hauteur de refoulement
to snore	schnarchen	ronfler
cut-out device	Ausschalter	coupe-circuit
body	Gehäuse	carter
volute	Spiralgehäuse	carter hélicoïdal
impeller	Pumpenlaufrad	rotor
brass	Messing	laiton
petrol engine	Benzinmotor	moteur à essence
impeller vane	Pumpenrad, Laufschaufel	ailette
plunger	(Tauch-)Kolben	piston plongeur
die-moulded	formgepreßt, profilgeformt	moulé sous pression
flap valve	Rückschlagventil	clapet
rod	Stange	barre
lever bar	Hebel, Pumpenschwengel	brimbale
gritty	sandig	sableux

English	German	French
sludge	Schlamm	boue
swivel base	Schwenk(grund)platte	plate-forme orientable
self-contained unit	komplette Einheit	groupe
pump body	Pumpenkörper	corps de pompe
hinged strainer	abklappbarer Saugkorb	crépine articulée
pump head	Pumpen-Oberteil	partie supérieure de la pompe
flexible drive	biegsame Welle	arbre flexible
power drive	Antriebswelle	arbre de transmission
plug -in coupling	Steck(kontakt)kupplung	boite de raccordement
quick release plug -in coupling	Schnell(steck)kupplung	raccordement rapide
ductile cast iron	Temperguß	fonte malléable
wire braid reinforced rubber	mit Drahtgeflecht bewehrter Gummi	caoutchouc renforcé par treillis métallique
thermo-switch	Thermo-(Schutz-)Schalter	commutateur à thermostat
casing	Gehäuse	carter
to connect in tandem	hintereinander schalten	installer en série
corrosive resistant	korrosionsfest	résistant à la corrosion
galvanised	verzinkt	galvanisé
corrugated	gewellt	ondulé
electric current	elektrischer Strom	courant électrique
feed cable	Anschlußkabel, Zuführungskabel	câble d'alimentation
rotary dewatering unit	Kreiselpumpen-(Grund-)Wasserabsenkungseinheit	groupe centrifuge d'épuisement
swampy	sumpfig	marécageux
excavation	Aushub	excavation
to cave in	einstürzen	ébouler
pit	Grube	fouille
clay	Ton	argile
marl	Mergel	marne
fissured rock	zerklüfteter Fels	roche à bancs fissurés
sump	(Pumpen-)Sumpf	puisard
shallow	nicht tief	peu profond
wellpoint system	Vakuumbrunnen-Anlage	système des pointes filtrantes
medium sand	Mittelsand	sable moyen
coarse sand	Grobsand	sable gros
shallow filter well system	Flach(filter)brunnen-Anlage	système des puits filtrants peu profond
screen	Sieb	tamis
jet	Düse, Strahl	ajutage
jetting	Einspülen	fonçage par jets d'eau
riser	Aufsatzrohr	tube d'aspiration
header pipe	Sammelleitung	collecteur
swing	Verbindungsleitung, Spiralschlauchbogenleitung	jonction pivotante
dewatering operation	(Grund-)Wasserabsenkung, Entwässerung	rabattement de nappe
to discharge	abfließen, ableiten, abführen	écouler
gate valve	Plattenschieber	vanne
cock	Hahn	robinet
inlet flange	Einlaßflansch	bride d'entrée
suction lock	Saug(anschluß)sperre	blocage d'aspiration
stepped switching system	abgestufte Schaltung	couplage gradué
jetting pump	Spülpumpe	pompe de lançage
ball valve	Kugel-Rückschlagventil	clapet à bille
suction	Saugen	aspiration

three-phase alternating current motor, rotary motor	Dreiphasen-Wechselstrommotor Drehstrom-Motor	moteur triphasé
standby pump	Reserve-Pumpe	pompe en réserve
failure, breakdown	Versagen, Ausfall, Panne, technische Störung	panne

Dialogue

Robert: Can you give me some idea, Duncan, how the requirements for dewatering system are calculated in Britain?

Duncan: Well, I would imagine our method differs very little, if at all, from yours on the Continent, but nevertheless, I shall be happy to oblige. The k-factor is the basis for the consultant's calculation, which can never be spot on, since the exact determination of this factor is seldom possible.

Robert: Before going any further, would you please explain this so called k-factor.

Duncan: Yes, certainly. Technically, it stands for the coefficient of permeability of a soil, which was conceived by Darcy. In less cumbersome language, it means basically the rate of flow of water at a given inclination in a soil. Incidentally, it is sometimes estimated from laboratory tests on undisturbed soil samples from site bore holes.

Robert: As you are well aware, such elaborate tests take up much time prior to the actual installation of a dewatering system. Because time is short in many cases, site investigation is often limited in practice to very quick soil tests, carried out just before the installation begins.

Duncan: It means, of course, that these rapid tests only produce very approximate results, and thus the sizing of the system can only be roughly assessed.

Robert: That's quite true, but in my opinion, this is all that's normally justified. In fact, with so many unpredicable factors, especially the k-factor, striving for a high degree of accuracy seems unnecessary.

Duncan: I agree with you entirely. It should be said, however, that laboratory tests do provide useful information for ground works generally, and not just for dewatering schemes. So when time permits and the job warrants it, I feel that a full site investigation is probably desirable.

Robert: Duncan, which type of dewatering system do you find most commonly adopted in Britain?

Duncan: I wouldn't like to say exactly, but as we do excavate in a great deal of clay here, the wellpoint system has met with a fair amount of favour, especially in connection with shallow excavations and trenches.

Robert: Isn't there a danger of the wellpoint filter becoming choked up with sludge?

Duncan: This may be so, Robert, but the problem of sludge can be overcome by surrounding each wellpoint with a graded filter, consisting of gravel and sand layers, to exclude fine particles from the wellpoint filter. Graded filters, as you know, may also be necessary in connection with filter wells. Incidentally, I believe you have some specialised knowledge of these filter wells, so perhaps you could tell our readers the main advantages and disadvantages of each type.

Robert: Yes, I've had some experience of them and will be only too pleased to do so. The shallow filter well system is only suitable for lowering the water table by about 5 to 7 metres, whereas in the deep filter well system the amount by which it can be lowered is limitless in theory.

Duncan: I think at this juncture, we should say that the extent and rapidity of the drop in the water table, depends very much on the performance of the submerged pump and the quality and size of the filter pipes etc.

Robert: A good point, Duncan, but let me add a little more information, if I may. In my view, the chief advantage of the deep system lies in its more reliable network – that is, all the pipes carry the water away under pressure and therefore absolute airtightness of pipe joints is not essential. With the shallow system, on the other hand, water is removed through suction, and here airtightness is vital, otherwise the network will break down.

Duncan: Yes, you're dead right. Shallow systems have actually been known to fail through as little as one faulty joint, and it can be very frustrating when there's a tight timetable to work to.

Robert: Surprising enough, I have even heard it alleged on occasions that the submerged pumps in a deep system are not altogether reliable, because they are not easily accessible for attention and inspection. Personally, I believe this is rather a myth.

Duncan: Well, we've already gained about 50 years experience with such pumps, and so I'm inclined to agree with you that this argument doesn't really hold water.

Robert: It's worth while, I'm sure, my saying a little about submersible pumps at this point. Their reliability depends, above all, on the motors and bearings, and it's interesting to note that these are literally surrounded in water.

Duncan: Excuse my interrupting you, Robert, but I would welcome the benefit of your experience in connection with the point you have just raised. Tell me frankly, are you really convinced that these water flooded motors – that is, with the water actually between the electric coils serving as a cooling medium, are not prone to short circuiting?

Robert: I must admit, it's logical to think that there might be some trouble, but in normal water this is not so. The improvement in the quality of recent insulation materials has made the motors virtually foolproof, even under hard working conditions. I would stress, however, that chemically affected water would most likely have detrimental effects on the insulation and it would, therefore, be advisable to employ pumps having their motors in separate chambers. In these types, only pure cooling water is actually in contact with the motor itself, which is isolated from the water being pumped.

Duncan: Many thanks indeed, Robert, for that useful bit of explanatory information. I wonder if I might persuade you to venture a little further whilst we're on the subject of submersible pumps.

Robert: Certainly, Duncan. There is one word of warning I'd like to put over concerning water lubricated plastic bearings, and that is not to run the motor for a minute or so out of water. I don't consider water lubrication to be at all unsatisfactory on account of this – don't get me wrong – but a little care is necessary.

Duncan: Can we now turn to something else of possible benefit to our readers – have you any ideas, Robert?

Robert: For a start, the old-fashioned mammoth pump, as it's generally known, is worth mentioning, to my mind. Successful systems using this form of pump are not unheard of even today.

Duncan: Yes, there's much truth in what you've just said about this old form of air lift pump. In cases where no convenient conventional pump is available on site, or where it is necessary to get rid of rather an abnormally large quantity of water, even when it is gritty, I would imagine that the rather primitive mammoth pump has its uses. I must confess, I've never actually seen one in use, so perhaps you would be good enough to explain exactly how it works.

Robert: It's really quite simple, Duncan. It is only necessary to lower a pipe into the water and force compressed air into its bottom through a smaller air pipe with a nozzle connected to its lower end. This nozzle is fastened to the bottom of the larger dewatering pipe to keep the air pipe steady. I can recall this principle being used many years ago in our coal mines.

Duncan: Yes, I can fully understand the idea now. It means, in fact, that provided enough air is forced into the riser pipe, the water column will be lifted. This is because its specific gravity, due to its increased air content, becomes less than that of the water outside the pipe.

Robert: To get away from specific points for a moment, I'd like to turn to the chief factor dictating the principle of all pumping systems.

Duncan: You must be referring to the atmospheric pressure of the air.

Robert: Yes I am, and perhaps we could devote a little of our discussion to this important factor. Atmospheric pressure, as you know, is about 1 kp/cm^2, and this allows us to lift pure water – I mean water without suspended matter – up to a theoretical height of 10 metres.

Duncan: The presence of atmospheric pressure leads us to the necessity for the suction pipe and casing of all pumps to be filled with water before running begins. Incidentally, this initial operation to induce the flow of water is better known as "priming the pump".

Robert: In the past, this operation often proved rather laborious, and for this reason most modern pumps are self-priming – in other words, they are equipped with a small air ejector, which draws air out of the pump prior to its working.

Duncan: Needless to say, the advent of the self-priming pump has been a most progressive step, since the old priming method first mentioned had repeatedly proved unreliable with many pumps, largely due to the non-return valve at the end of the suction pipe not being sufficiently watertight.

Robert: Talking of progress, one doesn't have to be very observant to notice the ever increasing use of pumping equipment purely for excavation purposes. I'm thinking mainly of coastal and river locations, for example. Have you any thoughts in this direction, Duncan?

Duncan: Well, as it happens, Robert, I have seen, on one or two occasions, pumps mounted on barges successfully engaged in moving immense quantities of sand from river beds in a relatively short time. I must say I was quite taken aback at the outstanding pumping capacity with such a material.

Robert: It may seem odd in this day and age, but this old method is still the principal one adopted for this particular operation, and indeed for other similar jobs. You spoke just now about immense quantities. Can you give more specific details?

Duncan: Well, I wasn't exactly on board the particular vessels I mentioned, so I'm unable to provide data from these actual operations. However, I do have some interesting information on the output of one of the biggest pumping devices employed in a new harbour construction on the Dutch coast.

Robert: Please fire away, I'm all ears.

Duncan: From a 1.00 metre diameter steel outlet pipe, the continuous volume of sand and water emitted was about 40 cubic metres per minute – quite astonishing. Incidentally, the volume ratio of sand to water was about 1:4 in this case.

Robert: The beauty of such a system is that the excavated material can be utilised at once to build a jetty or similar construction. Referring to the data you've just supplied, this would mean that enough material could be pumped out to provide a section 20 metres long, having a cross section of 700 square metres, in about one day.

Duncan: Yes, that's quite so, and furthermore the excavation itself could well have provided a suitable basin for shipping.

Robert: That's just the way in which the Dutch are constructing several of their harbours. Indeed, a simple and effective idea.

Duncan: Well, we started our discussion with dewatering systems and we've finished with harbour construction – certainly a wide diversity of topics. It's time to call a halt, I think, otherwise who knows where we might wander next.

Vocabulary	Wörterverzeichnis	Glossaire
k-factor, coefficient of permeability	k-Wert, Durchlässigkeitswert	coefficient de perméabilité
rate	Geschwindigkeit	vitesse
inclination, gradient	Längsneigung	pente
undisturbed soil sample	ungestörte Bodenprobe	éprouvette vierge
site investigation	Baugrunduntersuchung	reconnaissance du terrain
soil test	Bodenprüfung	essai du sol
ground works	Erdarbeiten und -bauten	terrassements
dewatering scheme	Entwässerungsplan	plan de drainage
graded filter	abgestufter (Sand/Kies-)Filter	filtre gradué
gravel layer	Kiesschicht	couche de gravier
network	(Rohr-)Leitungsnetz	réseau
pipe joint	Rohrverbindung	raccord de tube
to fail	versagen, ausfallen	tomber en panne
timetable, schedule	Zeitplan	programme
bearings	Lager	roulement
coil	Spule, Wicklung	bobine
cooling medium	Kühlmittel	agent refroidisseur
short circuit	Kurzschluß	court-circuit

insulation	Isolierung	isolement
foolproof	mißgriffsicher, betriebssicher	garanti contre toute possibilité d'erreur de commande
chamber	Kammer	chambre
plastic bearings	Kunststofflager	roulement plastique
mammoth pump	Mammut-Pumpe	pompe mammouth
nozzle	Düse, Austrittsöffnung	ajutage
riser pipe	Standrohr	tuyau vertical
suspended matter	Schwebestoffe	matières en suspension
non-return valve	Rückstauventil	clapet de retenue
barge	Lastkahn, Schute	chaland
vessel	Schiff	navire
data	Daten, Werte	données
volume ratio	Mischungsverhältnis nach Rauminhalt	rapport volume
jetty	Mole	mole, jetée, dique
basin for shipping	Hafenbecken	darse, bassin

(Reproduced by permission. Krafthand Verlag Walter Schulz)

(d) Air Compressors

Air compressors, which are used principally to power pneumatic tools as well as being employed in connection with such constructions as pneumatic caissons and tunnels, constitute quite a small, but nevertheless significant part of building plant. Rock drills, demolition hammers, concrete breakers, chipping hammers, scrapers, pile driving hammers, concrete vibrators, compactors, cement guns, riveting hammers, nut runners and scaling hammers are some of the common implements for which power is supplied through air lines from a compressor. Hoists, winches, chain blocks and concrete placers are also powered, on some occasions, by compressed air.

Compared with other modes of power, compressed air cannot be regarded as highly efficient mechanically, because it requires a comparatively large prime mover to produce an adequate supply of compressed air power. However, its overriding advantages are that it provides a transportable form of power, and that the many tools driven by it are relatively lightweight and simple to produce and maintain.

Basically, most compressors consist of a compressor unit, prime mover, being either an electric motor, diesel engine or petrol engine, oil reclaimer and cooling system and air receiver, all of which are usually completely enclosed by a sectionalised removable canopy. The canopy nowadays is often constructed of fibreglass and lined with polyurathene, thus reducing the overall weight of the machine and minimising engine noise. Pannier boxes for the basic tools are usually incorporated in the canopy.

In the vast majority of cases, the compressor is mounted on a portable chassis, having either four tyred wheels or two tyred wheels and an adjustable jockey wheel on the tow bar. Almost without exception, both these forms of chassis are provided with a towing hitch, which comprises an eye bolt and overrun brake operating mechanism on the end of the tow bar. Some of the smaller compressors are designed for mounting on a frame at the rear of a tractor and are easily demountable. The larger ones, intended to serve several tools simultaneously, are often truck- or crawler-mounted. When tractor- or crawler-mounted, a compressor has greater mobility, particulary on large undulating sites. The truck-mounted type has the advantage of being readily transportable at short notice from one site to another.

The rating of a compressor is governed by the volume of free air delivered in cubic metres per minutes (simply referred to as f. a. d.) at its maximum safe working pressure, being usually between 6 and 7 kp/cm^2. Alternatively, it can be rated rather more loosely as a one, two tool, or even larger machine, depending on the number of tools for which it is capable of supplying power.

Reciprocating (or piston) and rotary compressor units, both having their attributes, are those normally employed on most sites. Mainly due to its reliability and economic working, the well established reciprocating type is often still the more popular. However, the rotary type is sometimes preferred, because it has fewer component parts requiring maintenance, is less inclined to vibrate and is usually more compact and lighter in weight.

In general outline, compression with a reciprocating type is effected in stages by means of a differential piston. Low compression, the first stage, occurs on the up stroke and high compression, the second stage, on the down strokes. In this case, much of the heat of compression is extracted by a flow of cool air from a fan, and not by injected oil as in the rotary type. The temperature of compressed air is thus lowered and the volumetric efficiency, being the ratio of f. a. d. to piston displacement and usually expressed as a percentage, is conversely raised. As with the internal combustion engine, the more pistons that are employed the better the performance.

In the rotary type, the working principle differs greatly from that of the reciprocating type. One of these, the two-stage rotary vane unit, effects compression by means of two rotors in a stator. These are geared together in parallel, so that each runs at its most efficient peripheral speed in relation to its diameter. Through this arrangement, the air is allowed to flow in a smooth S pattern from intake to delivery. During this process, the sliding vanes of the rotors are subjected to the required centrifugal force to produce the maximum compression and an effective seal. From intake, the air flows from the low to the high pressure stage via the intermediate stage chamber, and is finally discharged into the air line. Throughout the compression process, oil is continuously injected through oil feed jets into both the low and high pressure rotor chambers, in order to (a) lubricate the moving parts, (b) absorb the heat generated and (c) thoroughly seal the sliding faces.

An interesting development in the rotary vane type compressor is the one which dispenses with the need of the conventional air receiver. In this particular variety, air delivery at a constant pressure is maintained by two features. One is a specially designed rotor having an increased number of sliding vanes, to give additional air chambers, and the other a servo valve.

Another rotary type is the oil-flooded single-stage rotary screw unit, which simply consists of two helical intermeshing rotors mounted inside a common casing. The male rotor has a number of lobes and the female rotor a number of flutes. The air from the inlet port flows into both the mating space between the lobes and the open flutes as the rotors revolve, and is thus compressed and subsequently discharged through the delivery port. Continuous injection of oil on the rotors ensures the lubrication of the working parts, the absorption of generated heat and the sealing of the working clearances.

Air delivery lines, usually of toughened rubber, are taken off the manifold of the compressor unit and each is fitted with an air cock. Once the normal delivered working pressure has been reached with a modern compressor, an automatic control device gradually reduces the speed of the prime mover. Any drop in the pressure is rectified by a pressure valve, which opens automatically, resulting in an increased speed of the engine. Silent working compressors are being used nowadays quite extensively, and many of the percussive tools they power are muffled to reduce noise.

"I think, Jack is a little off colour today"
(Reproduced by permission. Masters Agency U.S.A.)

Vocabulary	Wörterverzeichnis	Glossaire
air compressor, compressor unit	Drucklufterzeuger, Kompressor	compresseur d'air
pneumatic power tool	preßluftbetriebenes Werkzeug	outil pneumatiqué
pneumatic caisson	Druckluft-Senkkasten, Caisson	caisson d'air comprimé
rock drill	Bohrhammer	marteau perforateur
demolition hammer	Abbauhammer	marteau piqueur
concrete breaker	Beton-Aufbruchhammer	marteau brise-béton
chipping hammer	Meißelhammer, Absplitthammer	marteau burineur
scraper	Kratzer, Schabhammer	raclette
pile driving hammer	Rammhammer, Schnellschlagbär	marteau de battage
concrete vibrator	Beton-Rüttler	vibrateur à béton
compactor, punner	Stampfgerät, Verdichter	compacteur
cement gun, gunnite machine	Betonspritzmaschine	canon à ciment
riveting hammer	Niethammer, Nietrevolver	marteau-riveur
nut runner	Schlagschrauber	clé à choc
scaling hammer	Abklopfhammer	marteau détartreur
implement	Gerät, Werkzeug	outil
air line	Druckluftleitung	canalisation d'air comprimé
hoist	Aufzug	ascenseur de construction
winch	Winde	treuil
chain block	Kettenzug	palan à chaîne
earth drilling rig	Erdbohrgerät	engine de forage de sol
concrete pump	Betonpumpe	pompe à béton
pneumatically operated concrete placer	Druckluft-Betonförderer	transporteur pneumatique de béton
prime mover	Antriebsmaschine, Motor	mécanisme entraîneur
air power	Druckluftenergie	force d'air comprimé
oil reclaimer	Ölabscheider	déshuileur
air receiver	Windkessel, Druckluftkessel	réservoir d'air
sectionalised removable	in Teilen abnehmbar	amovible par éléments
canopy	Verkleidung, Abdeckung	capot
to line	beschichten, auskleiden	stratifier
overall weight	Gesamtgewicht	poids total
pannier boxes	seitlich angebrachte Behälter	bacs latéraux
tyred wheel	bereiftes Rad	roue à pneu
jockey wheel	Stützrad	roue de support
tow bar, draw bar	Deichsel, Zugstange	barre d'attelage
towing hitch	Anhänger-Kupplung	dispositif d'attelage de remorque
eye bolt	Zugöse	vis à oeillet
overrun brake operating mechanism	Auflaufbremse	frein à glissement
truck	Last(kraft)wagen, LKW	camion
crawler	Kettenfahrzeug	véhicule à chenilles
mounted	montiert, aufgebaut	monté
rating	Nennleistung	puissance nominale
free air delivered, f. a. d.	Luftliefermenge	rendement d'air
safe working pressure	höchstzulässiger (Arbeits) Druck	pression admissible
reciprocating compressor	Kolbenverdichter	compresseur à piston
rotary compressor unit	Drehkolbenverdichter	compresseur à piston rotatif
stage	Stufe	étage
differential piston	Stufenkolben	piston étagé
up stroke	Aufwärtshub	course ascendante
fan	Ventilator	ventilateur
to inject	einspritzen	injecter
volumetric efficiency	volumetrischer Wirkungsgrad	rendement volumétrique

piston displacement	Hubraum	cylindrée
vane	Flügelblatt, Lamelle	lame
geared together	durch Zahnräder verbunden	liaison par roues dentées
peripheral speed	Umfangsgeschwindigkeit	vitesse périphérique
S-pattern	S-Form	en forme de S
intake	Ansaugöffnung	orifice d'aspiration
delivery	Auslaß	échappement, sortie
seal	Abdichtung	étanchéité
feed jet	Einspritzdüse	injecteur
servo valve	Unterstützungs- oder Hilfsventil	servo-soupape
rotary screw unit	Schraubenverdichter	compresseur à hélice
helical	schraubenförmig	hélicoïdal
to intermesh	ineinandergreifen	engrener
casing	Gehäuse	carter
male rotor	Außengang-Rotor	rotor extérieur
female rotor	Innen(gewinde)gang-Rotor	rotor intérieur
lobe	Lappen, zahnförmige Leiste	crémaillère
flute	Rille	gorge
inlet port	Einlaßöffnung	orifice d'entrée
mating space	Paßraum	espace d'ajustage
delivery port	Auslaßöffnung	orifice de décharge
injection	Einspritzen	injection
clearance	Spiel, Toleranz	jeu
delivery line	Druck(luft)leitung	flexible d'air comprimé
toughened rubber	verstärkter Gummi	caoutchouc renforcé
manifold	Verteiler	collecteur
air cock	Druckluftventil, Drucklufthahn	robinet d'air comprimé
control device	Regler	régulateur
percussive tool	Schlagwerkzeug	outil frappeur
muffled	lärmgedämpft	insonorisé

Dialogue

Duncan: To kick off this discussion, Robert, perhaps you can please provide us with some fundamental data. I was thinking primarily about the performance requirements of a prime mover in relation to the volume and pressure of air delivered.

Robert: Yes, by all means. With modern equipment, an output of about 7 horse power (h.p) is usually necessary to produce 1 m^3 of delivered air per minute at a working pressure of 6 kp/cm^2.

Duncan: Those figures rather confirm that compressed air is not a particularly economic form of power.

Robert: That's fair comment, Duncan, for as a matter of fact it's the dearest. Mind you, it shouldn't be judged on this alone, for there are other considerations. Let's just look at one. As you are well aware, compressed air permits the use of relatively light weight tools, and this in itself is a major advantage.

Duncan: I entirely agree, because any labour saving factor these days is important and could possibly offset the high cost of producing the power. Another valid point in favour of compressed air, which shouldn't be overlooked, is that it is safe to use in underground work.

Robert: Yes, you've actually hit on the crux of the matter, in my opinion, and I'd like to enlarge on it, if I may.

Duncan: Please do.

Robert: In such situations as tunnels and mines, the use of electricity or internal combustion as a means of powering tools and equipment is seldom permitted. Compressed air is often the only feasible alternative, since any danger of explosion or air pollution is eliminated.

Duncan: Since we've mentioned underground work, it's worth adding that compressed air is also used extensively to improve the ventilation in such working conditions.

Robert: Don't forget, it cools the air at the working face at the same time.

Duncan: Yes, I must admit I had almost overlooked that point, but there is yet another feature which I haven't forgotten. An indispensible function of compressed air, as I see it, is to enable work to be done under water – in pneumatic caissons and shafts, for example.

Robert: I couldn't agree more, Duncan. Apparently, air compressed to about 3.0 kg/cm^2 above atmospheric pressure is sufficient to exclude water from such situations at a depth of less than 30 metres.

Duncan: Working shifts in compressed air at such a depth must be limited, of course, otherwise men in the working chamber will inevitably suffer from caisson disease, which is better known as the "bends".

Robert: Talking of underwater working, brings to mind another specific case, with which I was associated, where compressed air was the only suitable mode of power. In this instance, it was successfully utilised to operate a pile driving hammer device well below the water level under a bridge.

Duncan: I'm sure we have established a good case for compressed air, but in all fairness we should say something about its drawbacks.

Robert: Well, I suppose the foremost disadvantage is the noise it creates, which must limit its use somewhat. I often wonder though whether this really is such a serious fault these days. After all, many measures have already been taken to insulate the machinery and muffle the tools, especially the percussive ones like breakers and hammers.

Duncan: The noise element certainly isn't as objectionable as it was years ago – I'll grant you that – but the problem is by no means entirely solved. However, I believe we are in sight of a satisfactory solution.

Robert: I don't want to drop this topic yet, for I still have a few other points to make about the suppression of noise. Take the wider use of the helical screw type compressor, for example, which is quieter in operation than the others. This step and the introduction of silencers on both the suction and delivery lines have helped to reduce noise.

Duncan: This is quite so, and as a further measure to cut down the noise level, an insulation tent is sometimes erected to enclose the whole unit.

Robert: It's obvious that serious attempts have already been and are still being made to minimise any public annoyance and discomfort caused by compressors. Of course, a number of measures are enforceable by law.

Duncan: You mentioned discomfort, and so I think it's worth saying that an operative, in my view, suffers very little discomfort nowadays with up-to-date tools. Recently, I was asked by an operative to "have a go" with his new concrete breaker. I must say, I was staggered to find far less vibration than I had experienced some years ago with a similar tool. On the first occasion every bone in my body seemed to shake.

Robert: Before pressing on with other debatable items, can you think of any other noticeable drawbacks with compressed air?

Duncan: The size and weight of the machinery which has to be lugged around from place to place is the only other drawback that comes to mind.

Robert: This is always the criticism of self-contained power units. At the moment, I think we must accept this, in the hope that some revolutionary design will soon produce a small, compact and easily transportable unit.

Duncan: There's really little more I can add on this topic, so I'll ask you a question on something rather more specific. Which prime mover do you consider is best suited for the various types of compressor?

Robert: Well, it's generally recognised that the electric motor is the most suitable for the rotary compressor, because the rotary action is common to both. Likewise, the internal combustion engine is compatible with a reciprocating compressor, since the pistons act similarly.

Duncan: That seems perfectly logical to me. Now at this point, Robert, I'm sure you would like to talk about the term air receiver, which I know you're not too happy about.

Robert: Yes indeed. I think this is just another of your rather misleading terms – that is to a Continental – because, in fact, it is simply a compressed air container. Incidentally, it often forms part of the chassis on a wheel mounted machine.

Duncan: Agreed, it's a container, but I imagine the term receiver is used because the air is actually received by it after compression and prior to delivery. Anyway, don't lets make too much out of what might be a slight discrepancy in terminology.

Robert: By the way, talking of air receivers, it may be of some interest that it is advisable to release some pressure into them at the end of a working session. By doing so, the unit can be easily restarted on the resumption of work.

Duncan: Well, I think we've now covered the main issues, Robert, and I propose that we leave off in the hope that our readers have been given some food for thought.

Vocabulary	Wörterverzeichnis	Glossaire
to kick off	„losschießen", beginnen	démarrer
to offset	ausgleichen	balancer
underground	unterirdisch	en sous-sol
air pollution	Luftverschmutzung	pollution d'air
bends, caisson disease	Druckluftkrankheit	mal de l'air comprimé
to establish a good case	rechtfertigen	justifier
drawback	Nachteil, Schattenseite	revers de la médaille
silencer	Schalldämpfer	silencieux
noise level	Geräuschpegel	niveau sonore
operative	Arbeiter	ouvrier
to have a go	versuchen	essayer
self-contained power unit	Arbeitsmaschine mit Antrieb	machine indépendante
internal combustion engine	Verbrennungsmotor	moteur à combustion
piston	Kolben	piston

2. PILING AND BORING

Piled foundations constitute the most commonly applied system of providing a suitable base on ground of poor bearing capacity for heavy buildings and other important structures, like bridges, towers, quay walls, jetties etc. Broadly speaking, piles fall into two main categories, namely preformed and cast-in-situ. For preformed piles, reinforced concrete, tubular steel or sectional steel is normally employed, although timber is still sometimes used, especially for temporary structures. Hollow preformed piles are known as shell piles. The sinking of preformed piles into the ground is usually carried out with a piling plant.

Fig. 6 shows the well known "Menck" universal piling plant with a single-acting steam driven pile hammer at work. As can be seen, it is even possible to drive piles at a rake of about 45 degrees. This comparatively light, yet sturdy rig, constructed mainly of tubular frames, can also be equipped with other types of hammer. It is self-propelled, its superstructure slews a full circle and its winch machinery has drums for the pile, pile hammer and leader, as well as two capstan heads. Frame erection and all working functions can be carried out and controlled under the machine's own power. As is usual, the combined weight of the hammer and pile must not exceed the lifting capacity of the plant itself. There are similar, but larger rigs these days suitable for hammers with weights up to as much as 20,000 kgs.

The robust construction of a single-acting hammer, as used on the plant illustrated in Fig. 6, makes it particularly suitable for driving concrete and timber piles, sectional girders and sheet piling, even into hard strata. Because of the heavy blows inflicted on the pile, pile helmets are necessary in all cases. A helmet consists of a box-shaped steel cap fitting fairly closely over the head of the pile or double piling sheet, and it carries a hardwood packing on top, which must be renewed when hammered flat. Hammers and helmets are allowed to glide up or down the box-section leader by means of guides, which are controlled by a lifting cable connected to the hammer.

In any type of plant it is possible to employ a drop hammer. The simplest form consists of an iron block, known as a monkey, attached to a hoisting cable. To enable this heavy weight to strike the top of the pile, the cable drum of the steam, diesel or electrically operated winch must be released. In order to improve this striking action, the cable is often attached to a small weight, termed a cat, and not directly to the monkey. This cat should only be just heavy enough to allow the cable to glide along the leader. When coming into contact with the monkey, the cat hooks itself to it and both are lifted together by the winch. The monkey is able to fall freely down the leader, when the hooked coupling is released by pulling an attached rope. The cat and cable fall immediately afterwards, and the cat and monkey are thus automatically reconnected, ready to be lifted once more. With this system, up to 10 blows per minute can be struck on the pile head at a drop height of 1.25 metres. Nowadays, the simple drop hammer system has largely given way to the more efficient steam, diesel hydraulic, compressed air or electric hammer, which provides a much higher frequency of blows.

The insert of Fig. 6 shows the working principle of the popular Menck steam driven single-acting hammer. The rising cylinder of this particular model reverses automatically, by means of the control bar on top of the cylinder cover. It reverses in such a way, that the steam in the space above the piston, which causes the hammer cylinder to rise, can escape through the exhaust vent at the bottom of the cylinder. The upward movement is thus interrupted and the cylinder drops. In order to lift the cylinder again, the operator must actuate the control piston through the hand line and so admit a further supply of steam. The control bar is designed to allow the operation of the hammer at half or full stroke, and when only single light blows are required, the hammer can be manually operated. Whilst the cylinder acts as the striking element, the piston and piston rod remain stationary, but once the pile sinks into the ground, they follow behind by means of the hammer rail, which always remains in contact with the helmet and pile. The frequency of blows is normally 50 per minute, but this rate can vary according to the drop height of the hammer.

The weight of the hammer and the frequency of the blows must be adapted to suit firstly the ground conditions, and secondly the type and weight of the pile. In clay, loam and similar soils with a high degree of cohesion, piles require heavy blows at relatively long intervals, so that the pressurised water is able to

Fig. 6: MENCK & HAMBROCK Piling Plant

	tubular frame piling plant	Rohrgerüst-Ramme	sonnette à chevalet tubulaire
1	**cable sheave, cable pulley**	Seilrad, Seilrolle	poulie à câble
2	**leader**	Mäkler, Läuferrute	rail de guidage
3	**tubular steel member of tripod frame**	Stahlrohr des Dreigurtgerüsts	tube en acier du chevalet à trois montants
4	**working platform**	Arbeitspodest	plate-forme de travail
5	**chimney flue pipe**	Schornsteinrohr	cheminée
6	**vertical cross-tube steam boiler**	stehender Querrohr-Dampfkessel	chaudière à vapeur à tubulure verticale
7	**frame tilting spindle**	Gerüstneigungsspindel	broche d'inclinaison du chevalet
8	**winch machinery with cable drums**	Windwerk mit Seiltrommeln	treuil avec tambours à câble
9	**superstructure frame**	Rahmen des Oberwagens	châssis de la plate-forme
10	**driven axle**	angetriebene Achse	axe propulsif
11	**steam hose**	Dampfschlauch	tuyau à vapeur
12	**bull-gear ring**	Zahnkranz	couronne dentée

13	**girder of temporary bridge**	Hilfsbrückenträger	poutre de pont provisoire
14	**chassis**	Unterwagen-Rahmen	châssis du chariot
15	**rail**	Schiene	rail
16	**pile helmet, cushion head**	Rammhaube, Schlaghaube	casque
17	**hammer rail**	Bärschiene	montant de mouton
18	**pile hammer**	Rammbär	mouton
19	**control rope, hand line**	Steuerseil	câble de commande
20	**cranked pipeline**	Gelenkrohrleitung	conduite articulée
21	**pile**	Pfahl	pieu
22	**replaceable hardwood dolly**	auswechselbares Hartholzfutter	protection en bois dur amovible
23	**hammer rail**	Bärschiene	montant de mouton
24	**piston**	Kolben	piston
25	**leader**	Mäkler, Läuferrute	rail de guidage
26	**eyelet for hammer cable**	Öse für Bärseil	œil pour le câble du mouton
27	**control lever**	Steuerhebel	levier de commande
28	**steam feed**	Dampfeintritt	arrivée de la vapeur
29	**piston rod**	Kolbenstange	tige du piston
30	**control bar**	Steuerschiene	glissière de commande
31	**control piston**	Steuerkolben	piston de commande
32	**piston ports**	Kolbenkanäle	forages du piston
33	**hammer cylinder**	Bärzylinder	cylindre du mouton
34	**exhaust vent**	Auspuff	forage d'échappement
35	**pile helmet, cushion head**	Rammhaube, Schlaghaube	casque

flow away from the numerous voids between the thin flat particles of soil. In this way, the resistance of the soil around the pile is reduced. On the other hand, in sandy and gravelly soil, piles require lighter blows at much closer intervals, to prevent the granular and rounded particles of soil from hindering the movement of the pile.

Double-acting pile hammers, whether they are driven by steam, compressed air or other means, are more convenient in sandy or gravelly soil. This is so, because they can provide a frequency of light blows as much as ten times greater than that of a cylinder hammer. For the same reason, they are more suitable for driving steel sheets and sections in sandy soil than heavy concrete piles in cohesive soil. The frequency of the blows can be varied within certain limits, and thus an optimum efficiency can be achieved for each particular job. An important advantage of this rapid-stroke hammer lies in its ability to operate upside down to extract piles, but when used in this way, the kinetic energy developed is reduced to about one half of its potential. The control of double-acting hammers is fully automatic and their action can either be guided along the leader or free, as is the case when they hang directly over the pile on a cable from the boom of a crane. Up to the present day, this form of hammer, when driven by compressed air and provided with a watertight outlet, is the most effective for working under water.

The hydraulic hammer is similar in many respects to the compressed air or steam driven rapid-stroke type. The working principle is the same, except that it is fed with oil under pressure, which pulsates in a closed circuit. It is also fast working and can generally be employed in similar conditions to those suitable for other types of rapid-stroke hammer. The fact that it is less noisy, because it has no outlet, does give it some advantage.

Yet another form of hammer is the very simple and well proven diesel hammer. The oldest form, known as the "Dutch hammer", is still in use, mostly for marine structures. For many years it has been successfully employed driving piles and sheet walling in connection with protective measures against the destructive effects of crashing waves, and in the construction of quays for harbours etc. A modern form of this diesel hammer basically consists of a heavy falling cylinder, sliding between two hard chromium plated tubular side columns, and a fixed piston, the base plate of which is situated directly on the pile head. One column is ingeniously utilised as a fuel tank in additon to its normal function as one of the cylinder guide members.

To commence the operation of this diesel hammer, the cylinder is firstly lifted with the aid of a hoisting cable and then allowed to fall freely. At the end of its fall it strikes the piston base plate, compressing the air within the cylinder in the process. Just before the cylinder actually strikes this plate, a jet of diesel oil is injected through a hole in the piston into the combustion chamber containing the highly compressed air. As in the case of a diesel engine, the oil ignites, which immediately causes the air to become hot. On expansion of the air, the cylindrical hammer is forced upwards. When it reaches its highest point, having released a great deal of burnt gas, it falls down once again and the cycle repeats itself. The compressor, which pumps the oil into the highest part of the hammer, is a small yet robust pump, the piston of which is activated by a lever, which is directly pressed down by the falling cylinder. The extent of the fall can easily be regulated by a valve, which controls the quantity of fuel injected.

Despite the simplicity of this type, modified forms having a stationary cylinder and a heavy moving piston have proved more popular. It has been long established that the later type has a greater impact energy than the older version at the same low cost. The manufacturers of the efficient "Menck" diesel hammer, for example, consider that the improved performance is due to two reasons. Firstly, the weight of the hammer falling on to the anvil shaped block on the bottom of the cylinder provides part of the acting power, whilst the energy from expansion, which forces the piston up, also contributes in driving the pile downwards.

A quite different method of pile driving is fast becoming more evident nowadays, and this is by means of a vibro-pile hammer. This vibrating pile driver is particularly effective in granular soil, and on sites in thickly populated towns, it has the added advantage of usually conforming to the required suppression level for noise and vibration. It mainly consists of one or two electric motors with eccentric weights on their shafts, and in most cases has a mechanical excitation frequency of 50 cycles. The unit is connected to a power supply plant, providing between 35 and 250 kVA, via a switchgear trailer, and is normally built for 380 V. This type of hammer is able to operate freely on the end of a suspended cable from a crane boom, and the bottom part of the unit sits directly on the pile or sheet. Normally, the unit is friction connected by means of a built-on hydraulic clamping device, which enables it to extract sections as easily as it can drive them. In cases of extraction, a high pulling force must be applied. This is often achieved with a mechanically sprung extracting device hanging on the lifting cable and independent of the vibromotive forces produced by the unit.

A remarkable pile driving device originating in Britain and now being employed throughout the world is the "Taywood Pilemaster". It is relatively silent in operation, but has limited application. It is best suited to cohesive, yet not too solid soils, but unsuitable in strata of coarse sand and gravel. The decisive factor in the application of this device is that its dead load and the weight of the sheet walling, together with the reaction force created by the friction and cohesion of soil against the sheet walling, must exceed the resistance against driving in a double sheet. In the final process, the soil resistance is the critical force reaction.

Basically, the "Pilemaster" consists of a heavy cross head with 8 built-in hydraulic jacks beneath and the power plant on top. The power plant comprises an electric motor driving the hydraulic pressure pump, and provides added weight to press the wall of sheet piles into the ground. In order to sink a row of 8 piles, they are arranged and fixed in the required position, each piston rod being fastened to each sheet with bolts, plates and clamps. The device, which is freely suspended on a cable from a crane boom, is then allowed to exert its full weight of about 10 tonnes on to the top of the sheet pile wall section. The initial sinking of the section thus takes place. Only the two central jacks are now brought into action, and in suitable soil, the two central sheets are forced downwards 76 cm, which represents the full length of the stroke. Next, the two sheets on either side of those just sunk are also forced downwards, and afterwards the remaining sheets are similarly acted upon. With all the piston rods back inside their cylinders, this process, which is controlled by one man employing electro-magnetic valves, is repeated until the complete section is sunk to the required depth.

The alternative method to forming a piled foundation by driving piles or sheets is to bore holes to accommodate in-situ piles. This system is applied almost universally where noise and vibration must be avoided. Since there is such a variety of boring equipment, every type of soil can be penetrated by boring, especially as the action of drilling tools can often be assisted by water jetting or percussing when necessary. Boring is an old technique, which has been employed for a long time using simple tools and equipment to sink tubes for a well. Basically, the equipment consists of a drilling tool, such as an ordinary auger, continuous flight auger, sampling tube or drilling bucket, usually hanging on a cable from a pulley at the top of a tripod or from the boom of a crane or universal excavator. The other end of this cable is wound round the drum of a winch.

In normal conditions, simple boreholes are usually formed by sinking a strong steel tube into the ground. This is mainly achieved by removing the soil from within the tube, either by drilling it out or excavating it with a grab bucket. The lowering of the tube is assisted by its own weight and also by turning it by leverage. When boring to greater depths, further tube lengths of between 3 and 5 metres are added by threading them on to the preceeding lengths as they are sunk. In this way, it is possible to form really deep holes rather cheaply, because the equipment is quite simple and the labour force need rarely exceed 3 men. For added strength, or to combat the effects of aggressive ground water, the steel tube can be left in the ground, if considered necessary, as an outer casing to the concrete core. In most cases, however, the tube is withdrawn.

When using manually operated drilling tools, the actual type selected mainly depends on the nature of the strata. To get the best results, a little "trial and error" is often found necessary before the most suitable tool for the particular operation is finally decided upon. Both experience and time are required in such circumstances. To enable a suspended auger or spiral drill to be easily turned, it is provided with a square section shank to receive the two-handed lever bar. Once the operators know that the tool is full of soil, it is raised, set aside and cleaned before being lowered again into the tube.

A simple form of hand-operated auger can be made from a piece of pipe, which is cut lengthwise. One edge is hammered flat to produce a cutting blade when the tool is turned using an attached square section rod. Not all tools require to be turned to collect soil. For example, a sampling tube resembling an ordinary piece of pipe can be used effectively in muddy or sandy soil under water. From the end of a cable, or even attached to a round rod, such a tool is just dropped inside the borehole tube and then withdrawn to empty the contents. Sometimes such samplers have a closing device at the bottom to trap the soil, but in certain soil this is unnecessary. To penetrate hard strata, like rock, a heavy chisel type tool is necessary.

Although the somewhat primitive methods of boring just mentioned are economic in certain instances, they tend to be rather tiresome and time consuming in most cases. For many years, therefore, this operation has usually been carried out with special drilling rigs having power driven tools. With this equipment, the work is not only carried out far more quickly, but the hole can be cut more precisely, whether vertically or at an angle. The main parts of such a rig are the undercarriage, mounted either on crawler tracks, wheels or skids, the superstructure and the drilling frame. Forming part of the superstructure are the prime mover, all the mechanical devices for positioning and slewing the boring tool, and in modern equipment, the hydraulic pumps. The drilling frame comprises a steel mast, sometimes incorporating a leader to guide the tool and the motor, or gear, for turning the shaft.

Fig. 7 illustrates a "Wirth" all hydraulic universal drilling rig, which is well established throughout the world. This modern crawler-mounted rig with its swivelling platform is capable of boring large diameter holes with precision in all kinds of strata, and has been developed especially for foundation engineering. It is ideal when boring for piles, with or without casings, and for Berlin type retaining walls in difficult soil conditions. Amongst its features are an oscillator, which provides for casings up to about 1 metre in diameter, and a platform, which revolves a full circle to allow spoil to be easily discharged from a bucket or auger. The mast and oscillator can be tilted for boring at any angle up to 15° from the vertical. Through various combinations, the rig permits boring for piles or wells by different methods, such as jet suction, airlift, rotation or percussion. When using the direct circulation system, soil investigation or core cutting can be undertaken to depths of up to 600 metres. The B5R has diesel hydraulic drive with two hydraulic motors driving the power swivel, which can be swung sideways for free access to the borehole. The double

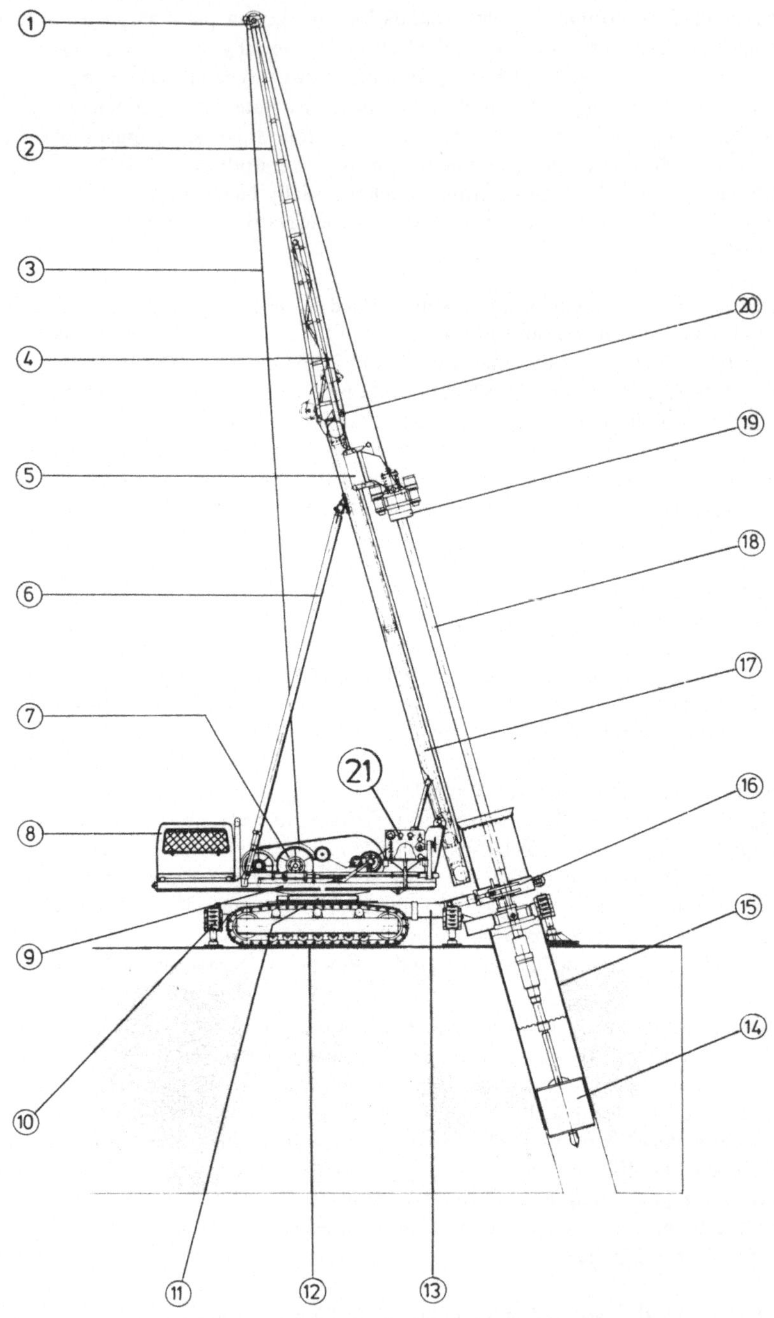

Fig. 7: WIRTH All-Hydraulic Universal Drilling Rig

1	**cable sheave, cable pulley**	Seilrolle, Seilrad	poulie de câble de rappel
2	**lattice extension mast**	Gittermastaufsatz	mât de prolongation en treillis
3	**carrying cable**	Tragseil	câble porteur
4	**erecting brace**	Aufrichter	bras releveur
5	**fixing clamp for power swivel**	Klemmbefestigung für Kraftdrehkopf	bride de fixation de la tête de transmission

6	**adjustable telescopic raking strut**	verstellbare Teleskopneigungsstütze	étai télescopique de réglage d'inclinaison
7	**cable drum of double winch drawworks**	Seiltrommel des Doppelwindwerks	tambours treuil
8	**air-cooled 8-cylinder diesel engine**	luftgekühlter Achtzylinder-Dieselmotor	moteur Diesel 8 cylindres à refroidissement à air
9	**superstructure**	Oberwagen	tourelle
10	**supporting jack**	Abstützpresse	stabilisateur
11	**bull-gear slewing ring**	Zahndrehkranz	couronne de rotation
12	**crawler track chassis**	Raupenketten-Unterwagen	châssis porteur à chenilles
13	**sliding beam for tilting adjustment**	Verschiebestück zur Neigungsverstellung	bras d'inclinaison réglable
14	**drilling bucket**	Bohreimer	garnissage de la tête de forage, seau de forage
15	**casing tube**	Verrohrung	tubage
16	**oscillator**	Schwinger, (Rohr-)Bewegungseinrichtung	oscillateur
17	**mast**	Mast	mât
18	**kelly bar**	Bohrstange	tige d'enbrasinement
19	**hydraulically driven power swivel**	hydraulisch angetriebene Dreheinrichtung, Kraftdrehknopf	tête rotative à commande hydraulique
20	**extension mast fixing bolts**	Verbindungsschrauben für Mastverlängerung	boulons de fixation pour la prolongation de mât
21	**operator's panel**	Armaturenbrett für Maschinisten	tableau de commande

winch drawworks with catheads are driven hydraulically with air-operated clutches. If required, a percussion device is available.

There is quite a variety of other types of drilling rig working on a similar principle and also capable of drilling the largest diameter holes. Some have an excavator as the basic machine, which is equipped with a special boom. On this boom, a hydrostatically driven rotary table is assembled about 7 metres above the ground. Other rigs are mounted on a lorry chassis, or even on a vessel, for easy mobility from one job to the next. With some equipment, it is possible to operate special rotary segmental grabs to excavate holes for the very largest diameter cast-in-situ piles, often with underreamed bases. In fact, all boring tasks are well within the scope of today's equipment.

When boring is adopted, the cast-in-situ piles are more usually of reinforced than plain concrete. The technique in employing reinforced piles only differs in detail from that used with plain piles. After the tube is sunk into the ground, the reinforcement cage, which is sometimes spot welded rather than fixed together with tying wire, is lowered inside the tube and placed in position. The concrete is then poured into the tube and packed around the cage, often by means of vibrators. Where a high water table exists and it is not intended to prevent water seepage into the tube, the concrete can be placed with a tremie.

There are numerous patented techniques for preventing water from entering the tube. One of these entails sealing off the top of the tube with an airtight cover, and forcing the water out by compressed air. With this process, the tube is forced upwards at the same time, and it is therefore advantageous to fill the hole with concrete in a plastic state, which can be tightly pressed against the earth sides of the hole. Alternatively, water can be kept out of the tube by sealing off its foot, either before or after it is sunk into the ground. In the first case, the foot is closed by an ordinary cap, and the sinking operation can only normally be done with a pile hammer. Where the foot is closed after sinking, a special patented closing device is employed.

Many of the patented in-situ concrete piles have an enlarged base, or foot as it is also known, usually in the form of a pearl, bulb or cone. The soil can be excavated to accommodate this base by special cutting blades on the end of a shaft. The blades are lowered into the tube in the folded position and opened out rather like an umbrella when reaching the bottom. They are then revolved to underream the hole and so form the enlarged excavation. In the case of some very large diameter boreholes for in-situ concrete piles, it is possible to manually form this enlarged excavation, since the workman can be adequately protected by the steel tubing.

The Franki piling system is often employed where in-situ concrete piles are required to carry extremely heavy loads. Its ability to do this has won it a worldwide reputation. In this system, the steel tube is firstly positioned vertically where the intended pile is to be sunk, and a bung of green concrete is placed inside the tube at the bottom. The concrete bung is then rammed into the ground with a cigar shaped hammer weighing about 2 tonnes. As the bungs sinks, the tube follows it, and once both have reached the required depth, the tube is lifted a little by two attached cables. Further concrete is placed in the tube and rammed, so that an enlarged foot is formed. The process of raising the tube a little, placing more concrete and ramming it is continued until the concrete reaches ground level. The shape of the pile depends on the resistance of the surrounding soil. Where the pile is reinforced, the cage is accurately formed and positioned to allow the hammer to pass easily within it. The Franki pile by its very nature is a combination of a cast in-situ and driven pile.

Vocabulary	Wörterverzeichnis	Glossaire
piling	Einbringen von Ramm- oder Ort-pfählen in den Boden	mise en œuvre des pieux
pile driving	Rammen von Pfählen	battage de pieux
boring, drilling	Bohren	forage
piled foundation	Pfahlgründung	fondation sur pieux
bearing capacity	Tragfähigkeit	capacité portante
structure	Bauwerk	ouvrage
quay wall	Kaimauer, Kaje	mur de quai
jetty	Mole	jetée
performed pile	Fertigpfahl, i. a. Rammpfahl	pieu préfabriqué
cast-in-situ pile, cast-in-place pile	Ortpfahl	pieu foré
steel pipe, tubular steel	Stahlrohr	tube en acier
timber	Bauholz, Nutzholz	bois de construction
hollow preformed pile	Hohl(fertig)pfahl	pieu foré creux
shell	Schale	coque
shell pile	Mantelpfahl	pieu à frottement
to sink	herunterbringen, abteufen	foncer
universal piling plant	Universal-Ramme	pilonneuse universelle
single-acting pile hammer	halbautomatischer Rammbär,	mouton à simple effet
rake	Neigung	inclinaison
rig, plant	Gerät	engin
tubular frame	Rohrrahmen	échafaudage tubulaire
hammer, pile hammer	Rammbär	mouton
self-propelled	selbstfahrend, mit eigenem Antrieb	automoteur, automotrice
superstructure, uppercarriage	Oberwagen	plate-forme
to slew, to revolve	herumdrehen, schwenken	tourner
winch machinery	Windwerk	treuil
drum	Trommel	tambour
leader	Mäkler, Läuferrute	glissière, rail de guidage

capstan	Spill	cabestan
frame erection	Aufrichtung des Gerüsts	montage de l'échafaudage
to control	steuern	commander
lifting capacity	Tragkraft	capacité
concrete pile	Betonpfahl	pieu en béton
timber pile	Holzpfahl	pieu en bois
sectional girder	Profilträger	poutre laminée
sheet piling	Spundwand	rideau de palplanches
stratum	Bodenschicht	couche de sous-sol
blow	(Ramm-)Schlag	coup
pile helmet, cushion head	Rammhaube	casque
piling sheet, sheet pile	Spundwandbohle	palplanche
box section	Kastenprofil	profilé en caisson
guide	(Gleitbacken-)Führung	guide
drop hammer, monkey	Freifallbär	mouton à déclic
hoisting cable, lifting cable	Hubseil	câble de levage
winch	Winde	treuil
cat	(Nachlauf-)Katze	ergot
hooked coupling	Klauenkupplung	embrayage à griffes
compressed air	Druckluft	air comprimé
frequency of blows	Schlaghäufigkeit	vitesse de frappe, nombre de coups
control bar	Steuerschiene	rail de commande
cylinder cover	Zylinderdeckel	couvercle de cylindre
exhaust vent	Auspuff	échappement
control piston	Steuerkolben	piston de commande
hand line	Abzugsleine	câble de rappel
stroke	Hub	course
manually operated	handbedient	commandé à main
to strike	schlagen	frapper
piston rod	Kolbenstange	barre de piston
hammer rail	Bärschiene	rail de mouton
ground condition	Bodenbeschaffenheit	conditions du sol
clay	Ton	argile
loam	Lehm	terre glaise
cohesion	Kohäsion	cohésion
pressurised water, water under pressure	Druckwasser, gespanntes Wasser	eau suspendue
void	Hohlraum	vide
particle	Teilchen	particule
gravelly	kieselig	graveleux
granular	körnig	granulaire
double-acting pile hammer, rapid-stroke hammer	Schnellschlagbär	marteau-trépideur
cohesive soil	bindiger Boden	sol cohésif
optimum efficiency	bester Wirkungsgrad	rendement optimum
to extract	(heraus)ziehen	arracher
kinetic energy	Bewegungsenergie	énergie cinétique
potential (energy)	Energie der Lage, hier: mögliche Energie	énergie potentielle
boom	Ausleger	flèche
crane	Kran	grue
outlet	Auslaß(öffnung)	orifice de décharge
hydraulic hammer	hydraulischer Bär	mouton hydraulique
to feed	speisen	alimenter
closed circuit	geschlossener Kreislauf	circuit fermé

marine structures	Seebauten	travaux maritimes
crashing wave	Brecher	brisant
harbour	Hafen	port
hard chromium plated	hartverchromt	chrome dur
tubular	röhrenförmig	tubulaire
column	Säule	colonne
base plate	Grundplatte	plaque de base
fuel tank	Kraftstofftank	réservoir de carburant
guide member	Führung(sglied)	guide
jet	Strahl	jet
to inject	einspritzen	injecter
combustion chamber	Verbrennungsraum	chambre de combustion
diesel engine	Diesel-Motor	moteur diesel
to ignite	entzünden	allumer
cycle	Takt	cycle
pump	Pumpe	pompe
lever	Hebel	levier
to regulate	regulieren, einstellen	régler
valve	Ventil	soupape
stationary cylinder	feststehender Zylinder	cylindre fixe
impact energy	Schlagenergie	énergie de frappe
performance	Leistung	rendement
anvil	Amboß	enclume
acting power	wirksame Energie	énergie efficace
to vibrate	schwingen	osciller
vibro-pile hammer, vibrating pile driver	Schwing- bzw. Rüttelrammgerät, Vibrationspfahlramme	vibro-sonnette
site	Baustelle	chantier
suppression level	Dämpfungsspiegel	niveau d'amortissement
eccentric weight	exzentrisches Gewicht	poids excentrique
shaft	Welle	arbre
mechanical excitation frequency	mechanische Erregungsfrequenz	fréquence d'excitation mécaniqu
(electrical) power supply plant	Stromaggregat, Stromerzeuger	groupe électrogéne
cycle	Periode	cycle
switchgear	Schalteinrichtung	équipement de distribution
trailer	Anhänger	remorque
suspended cable	Hängeseil	câble de suspension
friction	Reibung	frottement
built-on	angebaut	attaché
hydraulic clamping device	hydraulische Klemmvorrichtung	dispositif de serrage hydraulique
extraction	(Heraus-)Ziehen	arrachage
pulling force	Zugkraft	force de traction
mechanically sprung	mechanisch gefedert	suspendu par ressorts
vibromotive forces	Schwingkräfte	forces oscillatoires
coarse sand	Grobsand	gros sable
dead load	Eigengewicht	poids propre
soil resistance	Bodenwiderstand	résistance du sol
cross head	Querhaupt	traverse
hydraulic jack	hydraulische Hebevorrichtung	vérin hydraulique
bolt	Bolzen	boulon
plate	Platte	plaque
electro-magnetic valve	elektro-magnetisches Ventil	soupape électro-magnétique
boring equipment	Bohreinrichtung	équipment de forage
drilling tool, boring tool	Bohrwerkzeug, Bohrer	outil de forage
water jetting	Spülen(mit Wasser)	lancage
percussing	Schlagbohren	forage à percussion

well	Brunnen	puits
auger	Tellerbohrer	tarière à disque
continuous flight auger, spiral drill	Schraubenbohrer, Spiralbohrer	tarière à vis
sampling tube	Bodenentnahmerohr	tube unique, curette
drilling bucket	Bohreimer	seau de forage
pulley	Seilrolle	poulie
tripod	Dreibock, dreibeiniges Bohrgerüst	trépied
universal excavator	Universalbagger	pelle universelle
to excavate	ausheben, ausschachten	excaver
grab bucket	Greifer, Bohrgreifer	benne preneuse
leverage	Hebelkraft, Hebelanwendung	force d'un levier
tube length	Rohrschuß	élément de tuyau
to thread	aufschrauben	visser
labour force	Arbeitskräfte	main d'oeuvre
aggressive ground water	aggressives Grundwasser	eau phréatique agressive
outer casing	äußerer Mantel, Hülle	tube extérieur
concrete core	Betonkern	noyau en béton
"trial and error"	Experimentieren	essayer empiriquement
two-handed lever bar	Hebelstange für zwei Mann, Bohrkrückel	balancier
hand-operated auger, (clay cutter)	Bohrschappe	tarière
cutting blade	Schneide	tranchant
rod	Stab, Stange	barre
closing device	Verschluß	organe de fermeture
rock	Fels	roche
chisel	Meißel	trépan
drilling rig	Bohrgerät	équipement de forage
power driven	mit Motorkraft angetrieben	entraîné mécaniquement
undercarriage, chassis	Unterwagen, Fahrgestell	chariot
mounted	montiert, aufgebaut	monté
crawler track, caterpillar track	Raupenkette	chenille
drilling frame	Bohrgerüst	échafaudage de forage
prime mover	Antriebsmaschine, (Motor)	moteur
to position	in Stellung bringen	mettre en position
mast	Mast	mât
shaft	Welle, (Bohr-)Spindel	arbre
swivelling platform	schwenkbare (Arbeits-)Bühne	plate-forme tournante
foundation engineering	Grundbau (-technik)	technique des fondations
Berlin type retaining wall	Stützwand im "Berliner Verbau"	soutènement berlinois
oscillator	Schwinger, (Rohr-)Bewegungseinrichtung	engin oscillatoire
casing	Verrohrung	tubage
spoil	Aushub, Bohrgut	déblais
jet suction	Saugeinspülung	aspiration de lancage
airlift (method)	Lufthebe-Methode, Mammut-Pumpen Methode	méthode de pompe mammouth
soil investigation	Bodenuntersuchung	investigation du sol
core cutting	Kernbohren	carottage
diesel hydraulic drive	Diesel-hydraulischer Antrieb	entrainement diesel-hydraulique
power swivel	Kraftdrehkopf	tête orientable
cathead	Spill(kopf)	cabestan
double winch drawworks	doppeltes Windenzugwerk	treuil double
air-operated	druckluftbetätigt	commandé par air comprimé

clutch	Kupplung	embrayage
basic machine	Grundgerät	engin de base
rotary table, turntable	Drehtisch, Bohrtisch	table rotative
lorry	Lastwagen	camion
rotary segmental grab	rotierender Polypgreifer	benne polype rotative
underreamed base or foot	unterschnittener, erweiterter Fuß	base élargie
plain concrete	unbewehrter Beton	béton non-armé
reinforcement cage	Bewehrungskorb	cage d'armature
spot welded	punktgeschweißt	soudé par points
tying wire	Bindedraht	fil à ligature
vibrator	Rüttler	vibrateur
water table, groundwater level	Grundwasserspiegel	nappe phréatiquée
seepage	Durchsickern	percolation
tremie	Kontraktorbetonrohr	trémie
patented technique	patentiertes Verfahren	méthode brevetée
to seal off	fest verschließen, abdichten	étanchéiser
airtight cover	luftdichter Deckel	couvercle étanche
to underream, to undercut	unterschneiden	creuser
bung, stopper	Pfropfen	bouchon
green concrete	frischer Beton	béton frais
to ram	rammen	pilonner
ground level	Erdgleiche	niveau du terrain

Dialogue

Duncan: Looking at the fast development in piling techniques, do you think that the original timber piles stand any chance of survival against the more sophisticated piles?

Robert: Yes, I'm quite sure they do. Even today, they still prove to be the most economical for a number of purposes. Just think of underpinning and the many temporary structures, in which they are used with great success.

Duncan: Perhaps you would give some examples, Robert.

Robert: Certainly. One good example is scaffolding and framework for a pile driving rig to erect a jetty in the open sea, and another one is temporary foundations for exhibition platforms or buildings.

Duncan: Your last example brings to mind that rather extensive artificial island, which was created in Lake Geneva on the occasion of the Swiss National Exhibition in 1964. No such island exists now, and the many timber piles employed there were extracted soon after the exhibition closed. I'd like now to switch to steel piles. I understand that the Americans favoured them for quite some time, and I wonder what your thoughts might be.

Robert: That's true, but I think their preference was chiefly governed by economics. Whilst steel was comparatively cheap and always on hand, it was logical to make full use of it.

Duncan: Of course, an important factor with steel piles is maintenance, which largely means protection against rust.

Robert: I can hardly imagine anyone using a tubular steel pile, sheet steel piling, or a sectional steel pile, in aggressive water, for example, without giving it some protective treatment. For jetties or quays, which are subjected to the ill effects of salty sea water, the cathodic protection treatment is very helpful, as you probably know.

Duncan: To be honest, I'm not all that familiar with the precise details of this system, but I do understand the basic principle. To put it as simply as possible, one can say that an anode is provided, usually in the form of scrap iron, in order to prevent corrosion of the steel piles, which act as the cathode. As I see it, the anode, which is easily replaceable, is really a sacrifice for the benefit of the cathode.

Robert: Roughly speaking, that's correct, but it's necessary to add a couple of points. Where an iron anode is used, it must be connected to the positive of a direct electrical current, whilst the cathode is connected to the negative. An electrical circuit is set up in this way, with the salty water serving as a good electrolyte, and iron ions are transferred from the scrap iron to the surface of the piles. What it means, in fact, is that the anode is allowed to corrode, to provide the cathode with a protective coating. Cadmium can also be used as the anode, in which case no electrical energy is required, but iron is a more economical material and the electricity consumed is negligible.

Duncan: That's quite interesting, but according to the age old proverb "prevention is better than cure", it seems logical to use stainless rather than ordinary steel for piles. The only obstacle in the way appears to be the industry's inability, as yet, to produce stainless steel at a reasonable price.

Robert: In the construction industry as a whole, as far as I can see, only a very limited use has been made of pure stainless steel to date. I have, however, noticed a number of small bridges and electric pylons constructed with it, and I imagine the main motive was to save maintenance costs.

Duncan: For that very reason, it is easy to understand why reinforced concrete is often preferred to steel construction, and this is never more true than in the case of piles.

Robert: Now that we are back on piles, Duncan, maybe we can devote some time to the driven type.

Duncan: A good idea. Please go ahead.

Robert: In the Netherlands, for example, where driven concrete pile foundations are wide spread, they use both normal and prestressed reinforcement in the piles. Have you any views as to which is preferable?

Duncan: I would say without doubt that prestressed reinforced concrete piles are more suitable in all cases, provided that they are produced according to the stipulated standard and are driven properly. With both these requirements satisfied, no cracks will occur in the concrete cover to the reinforcing bars, which will thus remain protected against rust.

Robert: By the way, the belief that hammering destroys the bond between the reinforcing bars and the concrete is quite unfounded. On numerous occasions, later examination has shown that the bond remains firm.

Duncan: Tell me, what do you think about the idea, often expressed in old construction books, that the head and the foot of a concrete pile should be protected with steel?

Robert: As far as the foot is concerned, I think such protection is quite unnecessary, because of the high quality concrete used today. On the other hand, protection from heavy hammer blows is desirable for the head, and I feel this is satisfactorily achieved with a fitted steel helmet having an adequate shock absorbent packing material underneath.

Duncan: What do you class as an adequate material for this purpose?

Robert: This, of course, is dependent on many factors, such as the weight of the hammer, the frequency of blows and the type of pile. In cohesive soil, where heavy infrequent blows on the pile head are necessary to get the best results, a plastic material is useful to cushion the direct impact. An old practice, still being used I'll bet, was to shove an old piece of tyre between the pile head and helmet. It usually did the trick too.

Duncan: The use of a hardwood or plastic dolly placed in a recess on top of the helmet is the recognised cushioning method, is it not?

Robert: Yes, this is so, but it often helps to have a packing material as well. There are other essential considerations in pile driving, and an important one is the guidance of long piles to prevent buckling. The middle portion, at least, of a long pile must be held in the correct position at the commencement of hammering.

Duncan: I know, Robert, you've had first hand experience in the driving of long slender piles in the open sea. Could you pass on any particular observations of interest to our readers?

Robert: Indeed I can. A remarkable thing, which I have often witnessed in such a situation, is the assistance that the water gives in the driving of very long and slender piles. The stabilising effect it has is quite amazing, just as it is in underwater explosions. For example, if a reinforced concrete pile with a cross sectional size of 50 cm x 50 cm and a length of 25 metres is being driven at a commencing depth of 20 metres under water, there is no risk of damage to the pile, even when using an extremely heavy hammer.

Duncan: Incidentally, how do they manage to get such a long and slender pile under the helmet without breaking it? I've often wondered.

Robert: Well, Duncan, that's the crucial point in the entire operation of driving the pile. Before answering your specific question, it's best that I say a little about the design stage. The design of the longitudinal reinforcing bars by the specialist in statics is influenced more by the tricky handling operation in this case, than by the probable load or the hammering process. Now to answer your question. The pile is firstly hoisted by a crane from two lifting points, each located a fifth of the way along the pile from each end. To get the pile upright, the cable from the piling plant is fastened to a point one third of the way along the pile from its head. The cable drum of the winch is now turned and the foot of the pile is steadily guided along the ground, until the pile is in the required position.

Duncan: I think most of us realise that pile driving requires a lot of experience, and in many cases, comparatively expensive equipment is necessary. Of all the recognised methods, which one do you think is the least expensive?

Robert: Using a diesel hammer, I would say, both for concrete and steel piles. Compressed air or steam as the energy is far more expensive, even though each may prove the best method in particular instances. Some day perhaps, an electric driver will come on the scene, and I would imagine it might be similar to a vibro hammer, but somewhat cheaper.

Duncan: Up to the present day, the heavy steam hammer still remains the most suitable for driving large section piles in hard or cohesive ground. Turning to another aspect of pile driving, have you any comments about the attempts to muffle hammer noises?

Robert: Well, I've seen some odd attempts to suppress this noise, like wrapping the hammer with old curtains etc, but in most cases, I feel such a method is more of a psychological than a practical solution.

Duncan: You're probably right. Of course, the campaign to promote noise suppression is a general one, not just confined to hammers, and in my opinion, only revolutionary ideas can really overcome this dilemma. You see, one only has to think of all the many major sources of noise, such as the internal combustion engine in our millions of cars. In this example, we are also faced with the knotty problem of pollution of the environment, but maybe we shouldn't get involved in that at the moment.

Robert: I'm sure most of us look to electricity – electrically driven vehicles and site plant, for example – to solve this noise and pollution problem.

Duncan: Your optimism and leaning towards electricity are no surprise to me, and I agree that there are still tremendous possibilities, especially in respect of building plant. Well, we've strayed off course a little, Robert, so can you rectify this with any further interesting facts about noise suppression in pile driving?

Robert: Yes I can, Duncan. A striking example of endeavour to reduce noise to the very minimum occurred in the middle of Hamburg, in the construction of the underground tunnel for the city railway near the Alster.

Duncan: Wasn't that the sheet piling job which received criticism from some quarters, on the grounds that too much fuss was made to avoid any discomfort to the public and likely disturbance to the very old foundations of adjacent property?

Robert: Your're quite right, it was. Anyway, these efforts, whether they were overdone or not, entailed a lot of extra work. First of all, a deep trench was cut and filled with sand to replace the excavated soil, so that the sheet piling could be sunk therein almost without noise. The sheet piling forming a large pit was then sunk using a Taywood Pilemaster, which must be one of the least noisy of all pile drivers. The Benoto drilling equipment, which was also employed on this job to bore holes on the spot for "Secant" concrete piles, was by no means as quiet.

Duncan: To go to as much trouble as that to reduce inconvenience etc., naturally costs a great deal, but I suppose it can be justified if the equipment is advanced enough to cut down on manpower, let alone time.

Robert: As we have already spotlighted Hamburg, it might be as well to mention, that in the heart of this great city many deep pits have been planked and strutted using the Berlin method – that is with vertical soldier beams and horizontal timber planks. However, it's interesting to note that in recent years these soldier beams have not been driven by hammer, as had always been the practice. They have, in fact, been placed in open boreholes formed by sinking steel tubes, which are withdrawn later.

Duncan: Can we now talk about the means of overcoming stubborn resistance met by a pile foot or auger etc. in the occasional penetration of hard strata?

Robert: No doubt, you are thinking of jetting with water under pressure, but before we get on to that subject, what did you mean by occasional penetration?

Duncan: Well, I really mean that actual penetration would seldom be necessary, because any stubborn resistance met with in piling or boring usually suggests that a sound base has been reached, and there's no point in going any deeper. I realise, of course, that in some cases site investigation does reveal that at certain depths relatively thin strata must be penetrated, because they would be unable to carry the permissible load.

Robert: Now to discuss jetting. This is a very old method, particularly effective in strata like tightly compressed sand or loam. Anyway, despite its long history, practically all modern pile driving plants are equipped with water pumps and pipes for jetting.

Duncan: It's even possible, as you know, to have jetting pipes actually cast into reinforced concrete piles. To be of any real use, the water jet must be as sharp as possible and directed at the very point on the pile foot.

Robert: To get this really sharp jet, it has been proved in practice that the outlet area of the pressure pipe needs to be about four times smaller than the pipe itself.

Duncan: Mind you, a jetting device can also be used to pump other liquids, such as a thixotropic slurry in particular, into the base of deep foundations. This process helps to reduce the resistance to the movement of a pile or sheet, and at the same time makes a sheet pile wall, for example, watertight.

Robert: So much for jetting. I'd like now, in contrast, to turn to one of the more theoretical aspects of piling. We all know there are many formulae for calculating the loading capacity of a hammer driven pile. Most of them consider the normal factors, such as the weight of the hammer, the drop height and the weight of pile itself, but other factors are far more difficult to include in a formula. These less predictable factors are (a) the coefficients which characterise the resistance of the soil to be penetrated and (b) the acting energy, which is, in effect, the real kinetic energy, not only activated by a falling mass, but also by explosive pressure or a swinging mass.

Duncan: In varying soil conditions especially, I would think most formulae are worthless and often misleading.

Robert: I agree, and that's why testing is the best way, in my opinion, to determine the loading capacity. The unfortunate drawbacks of testing are the cost and time required, and time is usually a crucial factor at the foundation stage.

Duncan: There are, I believe, many testing methods. Robert, you are more conversant with these than I am, so maybe you could say a little about one of the more reliable ones.

Robert: The most reliable method is based on the principle of pressing and releasing a test pile by hydraulic jacks and a counterweight. The counterweight is usually in the form of heavy steel members, a block of concrete, or a large box filled with ballast. The upward and downward movements of the pile are recorded, and from a subsequent graph showing both the elastic and ultimate settlement, the specialist can find the answers relating to the permissible carrying load. This load is normally taken as half of the breaking load, which is indicated by a sharp bend in the load sinking curve at the moment it is reached.

Duncan: We've talked to a certain extent about the loading of piles, and I don't think we can really leave off without at least mentioning pile caps. These are, after all, distinctly connected with the distribution of the load.

Robert: Yes, you have a point there. A pile cap, in effect, is simply a heavy reinforced concrete slab cast around the tops of a group of piles, in order to unite them and evenly distribute the load from the structure amongst them.

Duncan: Before we wind up, there's just one other thing I'd like to mention about driving piles using both an outer tube, as in the Franki system, and an inner tube, termed a mandrel. A system, incidentally, which is used quite a bit in Britain. The idea is to drive the pile shoe into the ground together with in-situ concrete as it is placed. The shoe, usually comprising a hollow cast iron or steel concrete filled cone, is independent of the outer tube, but is connected to the reinforcing bars. Both tubes are struck by the hammer and the impact is transmitted to the metal perimeter and concrete core of the shoe. At the outset, the mandrel is filled with concrete, and once the shoe has reached the required depth, the mandrel is raised and replenished with further concrete. To form an enlarged foot to the pile, the outer tube is now raised and the mandrel is driven on its own. The raising and concrete filling of the mandrel, the raising of the outer tube and the driving of just the mandrel is repeated until concrete reaches ground level.

Robert: This idea, of course, is rather similar to the Franki system, except that a mandrel replaces the cigar shaped hammer. Well, I must admit we have only skirted around a number of topics relating to piling and boring, but it's almost impossible, to my mind, to give any real impression of the vastness of this subject. Reluctantly, I suggest we call a halt.

Duncan: Yes, if you like. We have not discussed such things as tunnel boring, drilling for injection anchors etc., I know, but not to worry, because I'm sure many of our readers will have whetted their appetite to find out a lot more on the subject as a whole.

Vocabulary	Wörterverzeichnis	Glossaire
to stand any chance, (to have any chance)	Aussicht auf Erfolg haben	avoir chance de succès
to underpin	abstützen, unterfangen	reprendre en sous-oeuvre
scaffold, scaffolding	Gerüst	échafaudage
framework	Rahmengerüst	bâti
platform	Plattform, Bühne, Podium	plate-forme
to switch to	(im Gespräch) übergehen zu	passer à
rust	Rost	rouille
sectional steel pile	Stahlprofilpfahl	pieu en acier
protective treatment	Schutzbehandlung	traitement protecteur
cathodic protection	kathodischer Schutz	protection cathodique
anode	Anode, positive Elektrode	anode
scrap iron	Eisenschrott	ferraille
cathode	Kathode, negative Elektrode	cathode
direct electrical current, (d.c)	Gleichstrom	courant continu
electrical circuit	elektrischer (Strom-)Kreis	circuit de courant
electrolyte	Elektrolyt	électrolyte
to corrode	korrodieren, zerfressen werden	être corrodé
protective coating	Schutzschicht	couche protectrice
stainless	nichtrostend	inoxydable
electric pylon	elektrischer Leitungsmast	poteau de ligne
maintenance costs	Unterhaltungskosten	coût d'entretien
prestressed reinforcement	Vorspannbewehrung	armature précontrainte
stipulated standard	Normvorschrift	règlement de normalisation
crack	Riß	fissure
concrete cover	Betonüberdeckung	recouvrement en béton
reinforcing bar	Bewehrungsstab	barre d'armature
to hammer	rammen	foncer par battage
bond	Verbund	liaison
shock absorbent	schlagmindernd, stoßdämpfend	amortissant
packing	Dichtung	garniture
to cushion	dämpfen	amortir
impact	Aufschlag	choc
to shove	schieben	pousser
tyre	Autoreifen	pneu

to do the trick, (to be successful)	den Zweck erreichen	atteindre le but
dolly	Rammhaubenfutter	garnitur
recess	Aussparung, (Kammer)	évidement
guidance	Führung	direction
buckling	Ausknicken	flambage
hammering	Rammen	battage
design	Entwurf, Bemessung	dimensionnement
to muffle	schalldämpfen durch ummanteln	insonoriser
noise suppression	Geräuschunterdrückung	lutte contre le bruit
internal combustion engine	Verbrennungsmotor	moteur à explosion
knotty	schwierig, verzwickt	difficile
pollution of the environment	Umweltverschmutzung	pollution de l'ambiance
building plant	Baumaschinen und -geräte	équipment des chantiers de construction
quarters	Teil der Öffentlichkeit	quartier
adjacent property	angrenzender Grundbesitz	propriété adjacente
to cut a trench	einen Graben ausheben	creuser une tranchée
pit	Baugrube	fouille
drilling equipment	Bohrausrüstung	équipement de forage
on the spot	an Ort und Stelle	sur place
to go to as much trouble as that, (to make as much effort as that)	sich soviel Mühe geben	se donner tant de mal
to cut down	herabsetzen, verringern	réduire, abaisser
manpower	Menschenkraft, Arbeitskraft	main d'œuvre
to let alone	nicht berühren	laisser seul
to spotlight	Aufmerksamkeit lenken	attirer l'attention sur
to plank	einbohlen	garnir de madriers
to strut	aussteifen	éntretoiser
soldier beam	Brustholz, Rammträger	étai vertical
timber plank	Holzbohle	madrier
permissible load	zulässige Belastung	charge admissible
water jet	Wasserstrahl	jet d'eau
outlet area	Austrittsfläche	surface de sortie
pressure pipe	Druckrohr	tuyau forcé
thixotropic slurry	thixotrope Spülung	boue thixotrope
loading capacity, bearing capacity	Tragkraft	capacité portante
drop height	Fallhöhe	hauteur de chute
hydraulic jack	hydraulische Presse	vérin
counterweight, counterbalance	Gegengewicht	contrepoids
ballast	Ballast	lest
ultimate settlement	bleibende Setzung	tassement final
breaking load	Bruchlast	charge de rupture
pile cap	Pfahlrostplatte	massif en béton armé solidarisant les têtes des pieux
reinforced concrete slab	Stahlbetonplatte	radier en béton armé
to cast concrete	betonieren	bétonner
to wind up, (to close or to end)	(Gespräch) beenden	terminer
mandrel	Dorn	mandrin
pile shoe	Pfahlschuh	sabot de pieu
in-situ concrete	Ortbeton	béton coulé en place
to place	einbauen	mettre en oeuvre
cast iron	Gußeisen	fonte

core	Kern	noyau
injection anchor	Injektionsanker	boulon injecté
at the outset, (at the beginning)	zu Beginn	au déput
to skirt around, (to go into little detail)	andeuten	esquisser

"May I ask where you studied engineering?"

(Reproduced by permission. Masters Agency U.S.A.)

3. CONCRETE TECHNOLOGY AND PRODUCTION

Concrete, by far the most utilised material in modern construction design, varies considerably in quality, depending mainly on its basic ingredients. These are normally cement, aggregates and water, which must all be carefully selected, accurately proportioned and thoroughly mixed, in order to produce a workable green concrete giving the required strength and density when hardened. The results from practice and research have enabled a high standard to be attained in this procedure nowadays. As opposed to timber and steel, for example, the properties of concrete can be largely determined by the site engineers themselves, who are able to exercise much control in its use.

A number of different cements are readily available to cater for the varying types of concrete structure and to combat the prevailing site conditions, such as the nature of the soil in direct contact with concrete foundations or ground beams. The correct cement for the particular situation can easily be determind by referring to the relevant standards. In Britain, for example, the British Standards gives comprehensive details of various cements, some of which are listed below as follows: –

(a) Ordinary Portland cement.
(b) Rapid hardening Portland cement.
(c) Portland blastfurnace cement.
(d) Sulphate resisting Portland cement.
(e) Low heat Portland cement.
(f) High alumina cement.

Aggregates, which are either natural or artificial in origin, are chemically inert filler materials, and must be structurally sound and durable, The aggregates for normal concrete should be sand, gravel or hard broken stone, free from dust, clay, loam, silt, or other impurities. For lightweight concrete they are usually crushed brick, blastfurnace slag, clinker, foamed lava, natural pumice stone, expanded burnt clay or shale.

For high strength concrete in particular, aggregates of the correct grain size composition must always be used, and these can easily be determined from the grading curves of sieve tests given under the relevant standard (e.g. DIN 1045 in Germany, or B.S. 882 1201 in Britain). The spaces, or zones, between the curves indicate the limits for varying qualities of aggregate, such as suitable and very good.

The water used should ideally be clean and free from practically all chemical reaction to achieve the best results, although water from natural sources, including sea water with some restrictions, is often acceptable. Of no lesser importance is the quantity of water added to the mixture. It can be generally said that a low water : cement ratio gives a strong dense concrete, whilst an excess of water produces a weak porous concrete when the surplus water has evaporated. The lowest limit for the water : cement ratio is largely controlled by the degree of workability and vibration necessary for the concrete. To get sufficient workability with a low water : cement ratio, it is necessary to restrict the quantity of sand used. This is particularly important in the case of fine sand, which absorbs much water. The bulking of sand on the addition of water is a factor which requires close attention. The gauging of the water content by weight rather than volume is to be recommended to ensure accuracy.

Good quality concrete depends a great deal on the following basic measures being implemented:

(a) a high cement rate,
(b) a low water : cement ratio,
(c) careful grading, with a high coarse aggregate rate,
(d) accurate batching of ingredients by weight,
(e) thorough mixing and compacting,
(f) sufficient curing.

Only for a large quantity of mass concrete, having sufficient strength by virtue of its bulk, can batching by volume be justified in certain instances. Crushing tests have proved conclusively that any former belief that the strength of concrete is largely controlled by the firmness of its aggregates is mainly unfounded. Such tests, however, have doubtless established that the weakest point of the conglomerate called concrete is the cement stone, which is, in fact, hardened cement paste. Nevertheless, the size and shape of aggregates is of great importance.

To improve or even alter the properties of concrete, admixtures, or additives as they are also termed, are often incorporated during the mixing process. The chemical industry has progressively produced a number of quite different additives for varying purposes, and some of the most common of these are given below as follows:

(a) Water with damp-proofing agents.
(b) Plasticizers.
(c) Accelerators.
(d) Retarders.
(e) Air-entraining agents.

Most additives are costly, and they must be proportioned exactly to prevent any adverse effect on other required properties of the concrete. For both reasons, their application should be limited to those instances where normal concrete mixes are inadequate for specific purposes.

Of the wide range of additives, the air-entraining agent has the most far-reaching application. Its popularity stems from two major features, one being that it increases the workability and so reduces the water: cement ratio of concrete, and the other that it improves the frost resistance of concrete. The air-entraining agent has been developed in the U.S.A. from certain resins or saponified oils. These form minute air bubbles, which work rather like ball bearings in the mixture to allow one particle to move against the other more easily, thus providing the green concrete with a degree of flexibility. Furthermore, the air-filled micro cells are particularly effective in acting as cushions against the pressure of ice from capillary water. Concrete used in good construction has a special call for an air-entraining agent.

The method of producing concrete varies considerably according to the quantity required. The production of really large quantities necessitates an installation made up of the following plant: –

(a) Aggregate bunkers or compartments.
(b) Cement silos.
(c) Batching equipment.
(d) Mixers.

Transportable plant is the normal type for most sites, whilst major construction sites and ready-mix concrete depots require static plant. Fig. 8 illustrates an ARBAU transportable concrete batching and mixing plant, which is self-contained and fully automatic. Its electro-pneumatic control of the aggregate batching and the electrical control of the cement and water batching, together with the up-to-date pan mixer, are well proven features. The speed of erection and compact design makes this plant very popular. The precise weighbatching of aggregates, which is an essential requirement for high quality concrete, makes it necessary to have separate stockpiles in their compartments of at least three grades of aggregate. Three of the most commonly stocked grades, for example, are 0–3 mm, 3–7 mm and 7–30 mm, all of which are available from natural sources, like river beds. Dredged sand and gravel from such a source are often covered with clay or other soil, and they must therefore be thoroughly washed and screened before being transported and deposited into the separated compartments.

As an alternative to natural gravel, crushed stone, which is often termed chippings, is equally suitable in most cases. In order to produce chippings, mechanical crushers are employed, and it often proves economical to operate a series of them to provide various grades from the same quarried material. The two principal types are the jaw and gyratory crusher. The jaw crusher is relatively simple in construction, merely having fixed and moving steel jaws, between which the stone is crushed. Very little power is required with such a machine. In the gyratory, or cone crusher, the idea is quite different, for in this type the stone is crushed between an eccentrically rotating conical head and a surrounding fixed bowl. Normally, the production rate from this type is higher than from the jaw type.

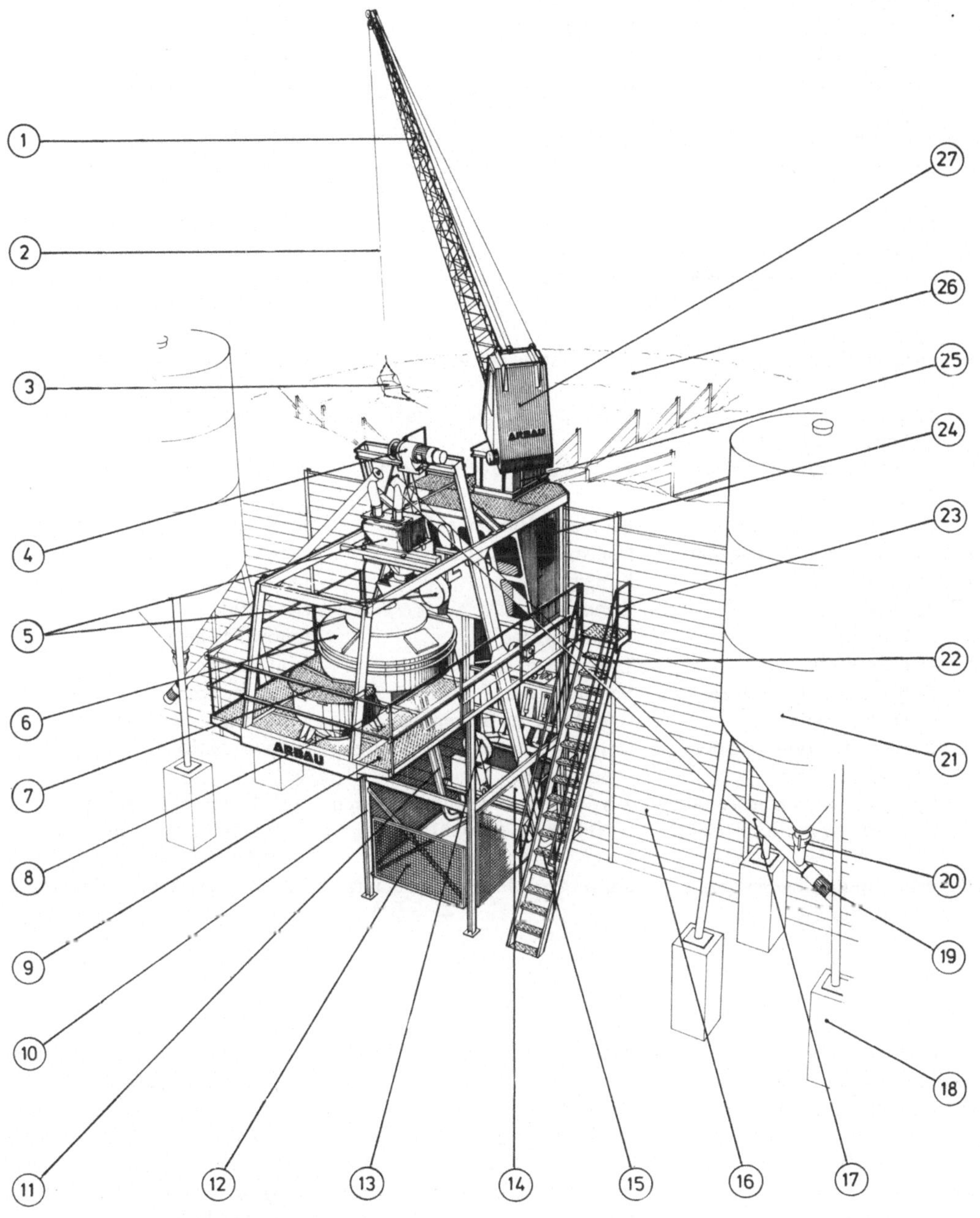

Fig. 8: ARBAU Self-Contained Concrete Batching and Mixing Plant

	self-contained concrete batching and mixing plant	Betondosier und -Mischanlage in kompakter Bauweise	installation compacte de dosage et de mélange de béton
1	**lattice jib or boom**	Gitterausleger	flèche en treillis
2	**steel cable**	Stahlseil	câble en acier
3	**dragline scraper**	Schrapperschaufel	benne de scraper
4	**geared lifting motor**	Aufzug-Getriebemotor	moteur de levage avec boîte
5	**cement weighbatcher with weighing dial head**	Zementwaage mit Gewichtsanzeiger	balance à ciment avec indicateur de poids

6	**pan mixer**	Teller-Zwangsmischer	malaxeur plat
7	**guard railing**	Schutzgeländer	garde-corps
8	**concrete outlet**	Betonauslaß	goulotte de sortie de béton
9	**chequer plate platform grating**	Gitterrost-Bühne	plate-forme en treillis
10	**steel framework**	Stahl-Rahmenkonstruktion	bâti support en acier
11	**guide track for skip**	Führungsschienen für Beschickerkasten	guides de la benne d'alimentation
12	**guard screen**	Schutzgitter	grille de protection
13	**aggregate batch meter**	Dosierungsanzeiger für Zuschläge	indicateur de dosage des granulats
14	**aggregate-weighbatcher skip**	Zuschlagaufzugskübel mit Chargenwiegevorrichtung	benne d'alimentation en granulats avec dispositif de pesage des charges
15	**chequer plate stair tread**	Treppenstufe aus Riffelblech	marches d'escalier en tôle cannelée
16	**horizontal timber planks between driven I-beams**	Bohlwand zwischen Rammträgern	paroi de madriers entre pieux battus
17	**steep-climb cement screw conveyor**	Zement-Steilförderschnecke	vis d'alimentation en ciment à forte pente
18	**concrete foundation base**	Betoneinzelfundament	fondation isolée en béton
19	**geared 3-phase electric motor**	Drehstrom-Getriebemotor	moteur à courant triphasé avec boîte
20	**coupling**	Kupplung	raccordement
21	**cement silo**	Zementsilo	silo à ciment
22	**stair string**	Treppenwange	limon d'escalier
23	**stair handrail**	Treppenhandlauf, Geländer	main courante
24	**fully electrically equipped control cabin**	Bedienungsstand mit vollelektrischer Steuerung	cabine de commande à équipement entièrement électrique
25	**base construction for boom dragline scraper**	Unterkonstruktion für Auslegerschrapper	support de flèche de scraper
26	**aggregates, separated according to size by partition walls**	Zuschläge nach Korngröße durch Teilwände getrennt	granulats entreposés dans des compartiments suivant la granulométrie
27	**cabin for dragline scraper operator**	Kanzel für Schrapperbedienung	cabine du conducteur du scraper

The screening of the crushed stone is either effected by vibratory steel screens or perforated rotating drums, which are slightly tilted. The stone is fed on to either one or the other screening device, in such a way that the entire quantity must pass over it. The movement of the screen, whether by vibration or rotation, allows the particles of varying sizes to drop through the appropriately sized apertures. Thus the stone is graded and falls directly into a hopper or chute, or alternatively, on to a conveyor belt or other means of transportation. Material mixed with clay or other impurity is washed either by a spraying or immersion process. In modern crushing plant, a rotary washer barrel is often combined with the grading device and is sometimes even mounted on a common chassis.

The storage of aggregates for a batching plant must be arranged, so that batching is possible either semi-automatically, through the manual operation of levers, or fully automatically, using electro-hydraulic or electro-pneumatic devices. This is normally effected either by having a row of bins or silos, under which discharging and weighing in the required proportions is carried out, or with a number of radiating compartments. These compartments are served by a common dragline scraper, which delivers the quantities of the various aggregates into the controlled feed-chute gates, prior to weighing.

In a small plant, a handscraper is usually employed to heap up the aggregates towards the discharge chute, and it is operated by the scraperman with the assistance of a winch. The scraper is, in fact, a shield and the action is similar to that of a plough. Larger plants nearly always utilise a boom scraper mounted centrally on top of the discharge chute. The movement of the scraper bucket is often controlled by an operator in a small cabin, or even automatically.

On batching plants, the cement is stored in silos from which it is drawn by means of a worm or helical conveyor, which must be absolutely dust-free. Where special circumstances dictate the use of bagged cement, a feeding hopper for this specific purpose is recommended. As with a silo, the cement can be fed from the hopper into the weighbatcher through a worm conveyor.

The chief part of a concrete production plant is, without doubt, the batching equipment combined with the mixer. Whenever possible, weighbatching of both aggregates and cement for normal concrete is usual nowadays and is, in any case, desirabe. For the better quality concretes, such procedure is called for by the concrete code of practice. The various aggregates necessary for the specific mix are normally weighed collectively. To do this, the outlet of each particular bin or compartement must be opened to discharge the required measures of sand, gravel and stone, until the single pointer on the large easily visible dial head reaches a predetermined dial reading. Because cement has a higher value than aggregates, it is always weighed separately, and the cement measure is indicated on a second dial head.

Several types of concrete mixer are in use at the present time, the principal ones being as follows: –

(a) Tilting drum mixer.
(b) Drum mixer with reversing discharge.
(c) Turbo mixer, known also as pan or paddle mixer.
(d) Constant-flow, or continuous mixer.

Both the tilting drum and reversing discharge drum mixers are gravity or free fall mixers, whilst the turbo and constant-flow mixers are pug mill mixers. The gravity and turbo mixers are generally referred to as periodical discharge, or batch mixers.

The gravity mixer is only suitable for dry mixing or for concrete mixes having particles of widely varying size. Such mixes only produce lumps, which merely change their shape, but never become thoroughly mixed. For stiffer concrete mixes, as are adopted in much modern design, mixers having agitating and raking arms are necessary. Alternatively, mixers with rotating drums and fixed arms can be used to produce much the same results. The shape and action of agitating parts have quite an influence on the quality of concrete. It is often possible to adapt the agitators according to the particular concrete mix required. Another developement in such equipment is in the use of replaceable paddles, which are elastically connected to the shaft to avoid breakages.

A pan mixer is normally regarded as a shallow drum which revolves in the horizontal plane, having a number of rotating paddles projecting down into it from a vertical spindle. Where the drum is stationary and only the paddles rotate, the mixer is then called a turbo mixer, which has been greatly developed in recent years.

Due to its reliability and high efficiency, principally on account of its short mixing times, the turbo mixer is now the most commonly employed mixer on a wide variety of sites. The mixing elements are usually made of high quality wear-resistant steel and the pan itself is often lined with wearing plates. Both these and the mixing elements are easily replaceable.The agitator is often provided with a contra-rotating impeller to improve the mixing process, and bottom and side scrapers in the pan guarantee thorough and intensive mixing. Some products have two sets of counter-running paddles as an alternative to an agitator. Concrete discharging is effected by means of one or even more segmental sliding plates operated either electrically, hydraulically or pneumatically. In the most modern forms, the water dosage is accurately measured by a water gauge and supplied through a quick-acting gate valve and water meter.

The expedience of concrete production is partly hinged on the number of working cycles possible per hour for a concrete mixer. A working cycle, in this case, is the time taken to charge the mixer with the ingredients, mix them and discharge the concrete. The actual mixing time, given by the relevant code of practice, is at least half a minute for a really efficient mixer and at least one minute for a mixer of average efficiency. The batching time, prior to mixing, is not considered in arriving at the working cycle time, since it normally coincides to a large extent with the other operations of producing concrete. Fully automatic plants having efficient mixers attain between 40 and 60 working cycles per hour.

Small batching plants are usually controlled manually and sometimes by push buttons. Mistakes in operation are not possible, because each successive step is tied in with the next. With larger modern fully-automatic plants, several selected mixing programmes, often as many as ten, can be fed to the equipment through programme cards, in much the same way that a fully-automatic washing machine is programmed to perform a specific job. Furthermore, the number of different mixes actually produced is often recorded, either by a counter or even in detailed printed form.

A constant flow of concrete is often desirable, as is the case where a continuously working pump is employed in transporting concrete. Here, intermediate concrete storage containers having their own agitators, or better still, continuous pug mill mixers are practical. The construction principle of such a mixer is derived from the well proven dough mixer, as used in the chemical industry to mix such items as paint or foodstuffs, for example. It consists basically of two slightly inclined half-round steel channels joined together, above which are fixed two staggered churning elements counteracting each other. The concrete ingredients are fed on to the highest end of the channels and churned over on route to the lowest end, well proportioned water being added. At the lower end, the concrete is ready for use and is thus discharged. By this method of constant feeding of ingredients, concrete is produced continuously. This procedure can be found working effectively and fully automatically on large static mixing plants all over the world, where huge quantities of concrete are required over a long period.

For the conveyance of concrete, pumping is often the most suitable and practical means. Alternatively, tower cranes using skips, conveyors, hoists, elevators and vehicles, like dump trucks and dumpers having tilting bowls, are also in common use. Vehicle bowls are sometimes provided with agitators, which can save valuable time during transportation.

The placing of concrete by pumping is no new idea, but the technique has changed over the years. The former method of using high capacity machines has mainly given way to the use of smaller capacity mobile pumps. The advantages of modern mobile pumps are many, but the principal ones are that concrete can be rapidly delivered to almost any required spot and their use is rarely curtailed by restricted working spaces, like in tunnels. No segregation of aggregates takes place, even over long distances, but on the contrary, the pumping action helps to complete the mixing process.

The huge standard two-cylinder reciprocating concrete pump, with its solid piston rods and crankshaft is now outdated. One of the main reasons for its inevitable loss of popularity was the everlasting annoyance caused by its mechanically controlled valves. Another dated device, still used even nowadays for conveying concrete through a pipeline, is known as a concrete placer. This consists of a pearl-shaped container with a hermetically sealed concrete inlet at the top and a delivery pipeline connected at the bottom. Once the container has been filled with concrete, compressed air is introduced on to the surface of the concrete, which is then ejected through the pipeline. At the end of the pipeline the concrete is discharged through a special discharge box into the shuttering. A major drawback of the concrete placer is that each charge of concrete is delivered with such intense force, rather like a rocket, which necessitates extremely solid jointing of the pipeline, especially at the bends. As a further precaution against the high impact, the pipeline must be firmly supported on trestles.

Mobile concrete pumps of modern design have already achieved world wide success. The majority of these chiefly consist of a two-cylinder pump, as with the older mechanical type, but with vastly improved pistons and valves, which are now driven more smoothly by oil hydraulics. To avoid damage through jammed pistons, the cylinders are flushed with water as an additional precaution. Fig. 9 shows a very efficient SCHWING mobile concrete pump. In this model, the pump and the folding distribution mast, similar to a crane boom, are mounted on the lorry, which houses the powerful prime mover. The mast and delivery pipe can both be folded hydraulically and slewed in a complete circle, so that concrete can be placed easily within a wide range and at a height exceeding 20 metres. Operation of the mast is simply by means of push buttons on a switch box on the vehicle or through a remote control push manual system, which enables the operator to direct it and control it at the pouring point.

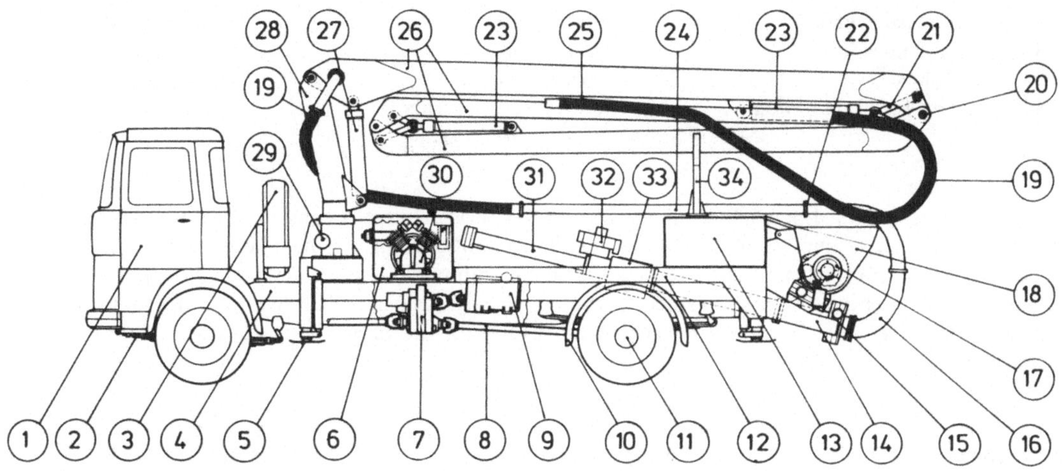

Fig 9: SCHWING Mobile Concrete Pump with Folding Placing Boom

	mobile concrete pump with folding placing boom	Auto-Betonpumpe mit knickbarem Verteilermast	pompe automobile à béton à mât articulé
1	**driver's cabin**	Fahrerhaus	cabine du conducteur
2	**leaf springs**	Blattfederpaket	ressort à lames
3	**spare wheel**	Reserverad	roue de secours
4	**chassis**	Fahrzeugrahmen	châssis
5	**hydraulic outrigger**	hydraulische Abstützung	stabilisateur hydraulique
6	**hydraulic tank**	hydraulischer Behälter	réservoir hydraulique
7	**transmission gear box**	Verteilerschaltgetriebe	boîte de transmission
8	**prop shaft with universal joints**	Gelenkwelle	arbre de transmission à cardans
9	**axial flow piston pump**	Axialkolbenpumpe	pompe axiale à piston
10	**mud guard**	Kotflügel	garde-boue
11	**wheel with tyre**	Rad mit Reifen	roue avec pneu
12	**ancillary frame**	Hilfsrahmen	châssis secondaire
13	**water tank**	Wasserbehälter	réservoir à eau
14	**pumping cylinder with control unit**	Förderzylinder mit Steuereinheit	cylindre de pompage avec organe de contrôle
15	**outlet control unit with gate valves**	Auslaßsteuereinheit mit Schiebern	organe de commande de sortie avec vannes
16	**bent tapering Y-junction**	Hosenrohrkrümmer, reduziert	tuyau de raccordement (échelle réduite)
17	**slip-on gear box**	Aufsteckgetriebe	boîte de commande amovible
18	**hopper with rubber border**	Trichter mit Gummirand	trémie avec bords en caoutchouc
19	**reinforced flexible concrete delivery hose**	verstärkter Betonförderschlauch	tuyau à béton renforcé
20	**hinged boom joint**	Auslegergelenkverbindung	articulation de mât
21	**guide lever**	Führungshebel	levier de commande
22	**coupling**	Kupplung	raccord
23	**folding cylinder**	Knickzylinder	cylindre articulé
24	**concrete delivery pipe**	Betonförder(rohr)leitung	conduite d'alimentation en béton
25	**flexible concrete delivery hose with nozzle**	Betonförderschlauch mit Auslaßstück	tuyau d'alimentation en béton avec bec de sortie
26	**boom section**	Auslegerabschnitte, Teilausleger	parties du mât
27	**lifting cylinder**	Hebezylinder	cylindre de levage du mât
28	**main column**	Hauptsäule	pied du mât
29	**slewing cylinder**	Schwenkzylinder	cylindre d'orientation
30	**compressor**	Kompressor	compresseur

31	**differential cylinder**	Differentialzylinder	cylindre de différentiel
32	**control block**	Steuerblock	organe de commande
33	**water box**	Wasserkasten	caisse à eau
34	**support for placing boom**	Stütze für Verteilermast	support de mât

Because of its unlimited possibilities, much research and development is constantly being carried out with concrete. Already, a number of substitutes for the conventional ingredients, like pulverised fuel ash and other by-products of industry, are well established, and improved additives are regularly appearing on the market. More attention has probably been given to lightweight concrete than to most other forms, but this is understandable, since it has a far reaching application for the future in practically all forms of construction.

Vocabulary	Wörterverzeichnis	Glossaire
concrete	Beton	béton
construction design	Bauentwurf	projet
basic ingredient	Grundbestandteil	élément de base
aggregates	Zuschlagstoffe, Zuschläge	granulat
workability	Verarbeitbarkeit	ouvrabilité
workable	verarbeitbar	ouvrable
green concrete	Frischbeton	béton frais
strength	Festigkeit	résistance
density	Dichte, Dichtigkeit	densité
research	Forschung	recherche
timber	Bauholz, Nutzholz	bois de construction
soil	Boden	sol
foundation	Fundament, Gründung	fondation
beam	Balken	poutre
ground beam	Streifenfundament, Bankett	fondation continue
blastfurnace cement	Hochofenzement, (HOZ)	ciment de haut-fourneau
sulphate resisting	sulfatbeständig	résistant aux sulfates
low heat Portland cement	Portlandzement mit geringer Abbindewärme	ciment portland avec faible chaleur d'hydratation
high alumina cement	Tonerdeschmelzzement	ciment alumineux
clay	Ton	argile
loam	Lehm	glaise
silt	Schluff	limon
slag	Schlacke	laitier de haut-fourneau
clinker	Kesselschlacke	mâchefer
foamed lava	porige Lavaschlacke	scorie volcanique
natural pumice stone	Naturbims	ponce naturelle
expanded burnt clay	Blähton	argile expansée
shale	Tonschiefer	schiste
grain size	Korngröße	calibre
grading curve of sieve test	Sieblinie	courbe granulométrique
water:cement ratio	Wasserzementwert	rapport ciment eau
cement rate	Zementanteil	pourcentage de ciment
to grade	(nach Korngrößen) abstufen	cribler
coarse aggregate rate	Anteil an Grobkorn	pourcentage en gros grain
batching	Dosierung, Abmessung	dosage
curing	Nachbehandlung	traitement ultérieur
crushing test	Druckfestigkeitsprüfung	essai de la résistance à la compression

admixture, agent, additive	Zusatzmittel	adjuvant
damp-proofing	Feuchtigkeitssperre	barrière anti-humidité
plasticizer	Betonverflüssiger	additif fluidifiant
accelerator	Erstarrungs- und Erhärtungsbeschleuniger	accélérateur
retarder	Erstarrungs- und Erhärtungsverzögerer	retardateur
air-entraining agent	luftporenbildender Zusatz	agent aérateur
resin	Harz	résine
saponified	verseift	saponifié
ball bearing	Kugellager	roulement à billes
batching equipment	Dosiereinrichtung	dispositif de dosage
mixer	Mischmaschine	mélangeur
transportable plant	Mobilanlage	installation mobile
site	Baustelle	chantier
ready-mix concrete	Transportbeton	béton préparé
static plant	stationäre Anlage	installation fixe
weighbatcher	Dosierwaage	doseur sur bande
weighbatching	gewichtsmäßige Dosierung, Wiegeabmessung	dosage par poids
stockpile	Lagerungshaufen, Halde	tas
natural source	natürliches Vorkommen	gisement
to dredge	naßbaggern	draguer
gravel	Kies	gravier
screened	gesiebt	criblé
crushed	gebrochen	concassé
crushed stone, broken stone	Schotter, Steinschlag	macadam
chippings	Splitt	gravillon
crusher	Steinbrecher	concasseur
jaw crusher	Backenbrecher	concasseur à mâchoires
gyratory crusher, cone crusher	Kreiselbrecher	concasseur giratoire
vibratory screen	Vibrationssieb	crible vibratoire
perforated rotating drum	Sieb(dreh)trommel	tambour cribleur
to feed	zuführen, beschicken	alimenter
screening device	Siebvorrichtung	dispositif de criblage
aperture	Öffnung, Loch	ouverture
graded	abgestuft, nach Korngröße	criblé
hopper	(Einlauf-) Trichter	entonnoir
chute	Rutsche	glissoir
conveyor belt	Förderband	transporteur à courroie
immersion process	Tauchverfahren	procédé d'immersion
rotary washer barrel	Wasch(dreh)trommel	tambour laveur
storage	Lagerung	stockage
lever	Hebel	levier
bin	Behälter	bac
radiating compartment	Fächerboxe	compartiment
served	beschickt	alimenté
dragline scraper	Schrapper	racleur
to control	steuern	commander
feed-chute	Beschickungsrinne, Einlaufrutsche	glissoir d'alimentation
gate	Verschluß(-Klappe oder -Schieber)	vanne
handscraper	Handschrapper	racleur à bras
to heap up	aufhäufen	amasser
discharge chute	Auslaßrinne	glissoir de décharge

winch	Winde	treuil
shield	Schild	écran
boom scraper	Auslegeschrapper	pelle de scrapage à flèche
mounted	montiert	monté
operator	Bedienungsmann	opérateur
to draw	abziehen	reprendre
worm conveyor, helical conveyor	Förderschnecke	vis transporteuse
bagged cement	Sackzement	ciment ensaché
feeding hopper	Beschickungstrichter	entonnoir
concrete code of practice	Betonrichtlinien	régles de béton
pointer	Zeiger	aiguille
dial	Zifferblatt	cadran
reading	Ablesung	lecture
tilting drum mixer	Kipptrommel-Mischer	bétonniére avec tambour basculeur
reversing discharge	Entleeren durch Umkehren der Drehrichtung	vidage par renversement du sens de la rotation
drum mixer with reversing discharge	Trommelmischer mit Umkehrgetriebe	bétonniére a contre-courant
turbo mixer, pan mixer, paddle mixer	Trog-Zwangsmischer	malaxeur horizontal
constant-flow mixer, continuous mixer	Durchlaufmischer	bétonniére continue
gravity mixer, free fall mixer	Freifallmischer	bétonniére à tambour inverseur
pug mill (mixer)	Rührwerkmischer	bétonniére-malaxeuse
batch mixer	Chargenmischer	mélangeur discontinu
agitator	Rührwerk, Nachmischer	agitateur
shaft	Welle	arbre
wear-resistant	verschleißfest	résistant à l'usure
to line	auskleiden	revêtir
wearing plate	Verschleißplatte	plaque d'usure
impeller	Wirbelradkranz	dispositif tourbillonant
scraper	Abstreicher	racloir
counter-running	gegenläufig	à marche contraire
water dosage	Wasserdosierung	dosage d'eau
water gauge	Wassermeßuhr	compteur d'eau
gate valve	Schütz, Schließventil, Schieber	vanne
working cycle	Arbeitsspiel	cycle de travail
code of practice	Richtlinien	règles
batching time	Dosierzeit	temps de dosage
push button	Druck(schalt)knopf	bouton poussoir
intermediate storage container	Zwischenbehälter	trémie intermédiaire
dough mixer	Knetmischer	malaxeur
channel	Rinne, Trog	goulotte
staggered	abwechselnd angeordnet	disposes alternativement
to churn	kneten	pétrir
to counteract	gegenläufig arbeiten	travailler dans le contraire
conveyance	Transport, Förderung	transport
skip	Kübel, "Bombe"	benne
conveyor	Materialförderer	convoyeur
elevator	Aufzug	monte-charge
dump truck	(LKW-)Kipper	camion basculant
dumper	Vorderkipper, Autoschütter	tombereau automoteur
tilting bowl	Kippmulde	benne basculant en avant

segregation	Entmischung	ségrégation
reciprocating pump	Kolbenpumpe	pompe à piston
piston rod	Kolbenstange	tige de piston
crankshaft	Kurbelwelle	vilebrequin
valve	Ventil, Schieber	vanne
concrete placer	Beton-Druckluftbeförderer	transporteur de béton pneumatique
hermetically sealed	luftdicht abgeschlossen	étanche à l'air
inlet	Einlaß(öffnung)	ouverture d'entrée
delivery pipeline	Förderrohrleitung	tuyauterie de décharge
discharge box	Fangkessel	réservoir d'arrivée du béton
shuttering, formwork	Schalung	coffrage
bend	Bogen	coude
trestle	(Stütz-)Bock	tréteau
mobile	beweglich, auf Rädern, motorisiert	mobile
to flush	durchspülen	chasser
folding distribution mast	knickbarer Verteilermast	mât de distributation articulé
crane boom	Kranausleger	flèche de grue
lorry	Lastwagen, (LKW)	camion
prime mover	Antriebsmaschine, Motor	dispositif entraîneur
to slew	herumdrehen, schwenken	tourner
switch box	Schaltkasten	armoire électrique
control push manual system	Drucktastensteuerung	commande par bouton-poussoir
to pour	betonieren, schütten	bétonner
pouring point	Betonierstelle, Aufgabestelle	point de coulage
pulverised fuel ash	Schlackenmehl	cendre volante
lightweight concrete	Leichtbeton	béton léger

Dialogue

Duncan: As is generally known, important progress has been made in the field of concrete technology Of all the notable achievements, which one do you think is the most outstanding?

Robert: Together with the interesting development of compacting tools, like vibrating pokers etc., and other useful machines, the ability to pour green concrete in a dryer state than was hitherto possible is, in my opinion, a really marked step forward.

Duncan: This inevitably leads me to ask you, Robert, to what extent you reduce the water: cement ratio on the Continent to achieve this?

Robert: It all depends on a number of factors, Duncan, such as the aggregate composition and, to a larger extent, the size of the concrete member in question. Generally speaking, it's possible to reduce the ratio up to a point where the concrete still remains workable enough.

Duncan: Now I have another leading question. What exactly do you mean by workable, and furthermore can you give some data regarding water:cement ratio?

Robert: In this particular sense, workable means completely compactable in the normal way. Now to give you some figures relating to water:cement ratio. Experience has shown that the highest concrete pressure strength of a mix having a convenient grading of the aggregates and 300 kgs of cement per cubic metre of set concrete is obtained when the water:cement ratio is between 0.4 and 0.5.

Duncan: That's quite interesting. At this point it would be as well to mention that a number of institutes, amongst them the American Concrete Institute, or A.C.I. as it is better known, produced a table in the early days, from which the maximum advisable water:cement ratio could be found for all kinds of concrete member under different watering conditions and in varying temperatures.

Robert: Yes, and such a table is invaluable, believe me. Anyway, in closing this important topic, I cannot refrain from stressing, that from two comparatively good workable mixes, the one with the smallest water:cement ratio gives the higher strength.

Duncan: Let's not forget to add that this fact has nothing to do with one having a greater cement content than the other.

Robert: Quite so. Since we are now dealing with the strength of concrete, it's also necessary to say that the bigger the aggregate size, the stronger the concrete.

Duncan: Yes, that's perfectly true, because larger chippings or pieces of gravel have comparatively smaller surface areas to be wetted than sandy aggregates.

Robert: You see, Duncan, we are again back to the vital water:cement ratio factor, which is reduced when large aggregates are used. Of course, in members having small dimensions or reinforcement, the maximum size of the aggregate is often limited.

Duncan: How much importance do you attach to the curing of concrete on the Continent, Robert? As a matter of interest, we in Britain generally consider that the watering of set concrete for a fair period is quite an important operation.

Robert: I am convinced it's very important. Some years ago I was most impressed by a chart showing the results of pressure strength tests over a long period on concrete samples which had received varying water treatments. The chart was, in fact, produced by the Portland Cement Association.

Duncan: Can you recall any interesting facts revealed by that chart?

Robert: I can indeed, Duncan. In a nutshell, it proved conclusively that concrete which has been laying fairly wet for 7 days will only strengthen considerably over a period of 5 years, for example, if the atmosphere is moist. In other words, left in a dry atmosphere for 5 years, such concrete would hardly improve at all.

Duncan: One can clearly see from this why the curing of concrete in roads is so important. The higher density resulting from proper curing resists the damaging effects of frost.

Robert: For the same purpose, air-entraining agents are employed in concrete roads, and good results are possible, although it is at the expense of a little compressive, as well as bending tension strength.

Duncan: By the way, how can you check these strengths?

Robert: The normal, and still the best way incidentally, to determine the compressive strength is by loading a standard concrete beam with one or even two loads as per the relevant code of practice. In both cases the breaking load must be ascertained, and from this the permissible, or safe, and breaking stresses are calculated.

Duncan: Just as we do, Robert, and perhaps I could add a little more detail to what you have said.

Robert: Please do.

Duncan: Concrete cubes, usually being about 100 or 150 mm thick, are cast on site in moulds, cured in normal conditions for 7 or 28 days, as required, and crushed. The permissible stresses are calculated as follows: –

0.19 × the 28-day strength for direct compression, and

0.25 × the 28-day strength for compressive stress in bending.

Robert: Where there's some doubt, another method of testing the strength of hardened concrete, known as the dynamic ball impact test method, is used here in preference.

Duncan: Yes, we are also familiar with this test, but I'm not too happy about the accuracy of the results, since they only show the hardness on the surface of the concrete. Furthermore, the results do depend largely on the skill of the tester.

Robert: Tell me, Duncan, are cylinder tests employed in Britain as they are in the U.S.A. or France?

Duncan: Not often to my knowledge. By the way, would you please enlarge a little on cylinder tests.

Robert: Certainly. The tests are made on hardened concrete cylinders about 150 mm in diameter and 300 mm long. For geometrical reasons the results are about 25 % poorer than for the same concrete crushed as a cube.

Duncan: Quite an absorbing topic, but maybe we could now move on to the next. Since you suffer worse winter conditions than we do here in Britain, I would be interested to hear your views about the placing of concrete in wintertime.

Robert: The very thought of placing concrete, or performing any other construction operation in fact, in freezing weather makes me shiver, but of course, work can't come to a halt and the authorities and contractors naturally want to continue working.

Duncan: I remember some years ago the German government was particularly generous in handing out subsidies and credit facilities at low interest rates to contractors for winter construction equipment.

Robert: That's quite true, and as a further incentive to keep the wheels of the industry moving, they even paid bonuses to workmen who were willing to work in bad weather conditions.

Duncan: This seems to me to be a sensible and commendable policy, which encourages production, and at the same time, saves the cost of unemployment benefits to workmen who would otherwise be laid off work. Before I find myself drifting into politics, I'll get back to the point of placing concrete in low temperatures. At what temperature on the Continent do you stop laying concrete in the normal way – that is without any protection or similar precautionary measures?

Robert: Actually, there's no serious problem, for it's relatively easy to place concrete at air temperatures down to about −4°C. One way to make this possible is to introduce steam into the aggregate heaps by means of pipes. In addition to this, of course, heated water can be used in the mix, but care must be taken not to heat the water beyond 80°C. Above this temperature, hot water in contact with cement could cause a flash set, which means the unwanted rapid setting of cement.

Duncan: Robert, what's your reaction to the use of antifreeze admixtures in concrete during cold spells?

Robert: You're obviously referring to those salts – mostly, chlorides or carbonates – which bring the freezing point down and accelerate the hardening process. As you know, Duncan, one must be particularly guarded in the use of chlorides in concrete. Reinforcement or embedded steel installation pipes, for example, will be attacked by rust corrosion caused by these chlorides.

Duncan: It's as well to mention another disadvantage when these frost protection agents are used liberally, namely that they can produce rather unsightly efflorescence at a later date. On the other hand, of course, it's quite useless to add to the mix only minute quantities of calcium or sodium chloride – the common term of the latter being table salt, by the way. This will have the contrary effect of prolonging the hardening process.

Robert: Speaking of such chemicals, could you say something about lignosulphonate, which I believe is more commonly used in Britain than here on the Continent.

Duncan: Lignosulphonate is, in fact, one of the basic ingredients of several admixtures. In it's varying forms it is suitable for a number of purposes, the main one being to retard the setting of concrete.

Robert: It's a mistaken belief that admixtures can perform miracles – that is, turn a poor concrete into something out of this world. Only the proper blending of all the correct concrete ingredients, including the admixtures, can produce the desired results. Now to move on to something else, Duncan, perhaps we should attempt to clarify the meaning of a number of general terms relating to concrete, which are often misunderstood.

Duncan: Yes, I'm all for that. Of the many commonly used terms, the ones that I have seen misused time and time again are "consistence", "plastic mix" and "workability". Explanations for these are more in your province than mine, so I'll pass the buck back to you.

Robert: Well, I'll certainly try to get it over to our readers as best I can, and hope you'll agree with my rather practical explanations. "Consistence" refers to the degree of deformability of all concrete mixes from the driest to the wettest – in other words, the degree of stiffness. A "plastic mix" is one which holds together without disintegrating and thus floats smoothly without demixing itself. Finally, we come to "workability", which means the ease with which concrete is poured into position. Workability naturally depends very much on the consistence of the concrete and also on the width and reinforcement of the unit being cast.

Duncan: That's jolly good, Robert. I'm sure your definitions will be fully understood. Now to the next point. I understand that in your country, and here in Britain to my knowledge, there is still no limit of practical value for concrete consistence and workability.

Robert: Yes, this is regrettably true, and I often wonder why this is so.

Duncan: I can't offer you any good reason for such an omission, I'm afraid, but I do feel, as I know you do, that compacting factor and slump tests have only a limited value. Take the slump test, for example, as employed both on the Continent and here in Britain. The primitive equipment used in carrying it out, namely a flow table and conical mould, leaves a lot to be desired in its accuracy. However, both tests do make it possible to roughly compare the consistence and workability of two specific mixes.

Robert: You'll no doubt go along with the idea of discussing the pumping of concrete next, since this process is employed a great deal on the Continent. With regard to the concrete mix, for example, what requirements do you consider are most important?

Duncan: For a start, the mix must contain at least 50 % sand containing a certain amount of ultra fine material, and it is always advisable to provide the delivery pipe with a smooth lining at the outset. This is done by pumping first of all with a mix rich in cement content. Other useful measures for pumping are the use of rounded aggregate, where possible, rather than sharp chippings to assist the flow of concrete and the incorporation of a plasticiser in the mix. This is to reduce the surface tension of the water content, and thus make it more liquid.

Robert: May I chip in here, whilst I think of it, just to mention how electrically driven vibrators are being used nowadays to assist in pumping concrete.

Duncan: In what way, exactly?

Robert: They can be actually welded directly on to any narrow bends in the delivery pipe, so as to prevent the concrete from clogging up at these points.

Duncan: Since we are discussing the pumping of concrete at the moment, we mustn't overlook the reliable and well proven type of pump, known as the "squeez-crete" concrete pump. The principle is quite different to that of the conventional piston pump, in that the concrete is squeezed along an elastic hose by the action of rubber rollers pressing down on the hose.

Robert: Rather like toothpaste being squeezed out of its tube.

Duncan: Yes, that's rather an apt description. The advantages of this system, incidentally, are that it's so simple in operation and it enables the movement of concrete to be reversed, should a blockage occur.

Robert: Reverting back to mixing plant, I think the idea of a skip also acting as a pug mill mixer is a good one. It certainly shortens the time cycle.

Duncan: This idea is used in what is known as the Mix Mobil, isn't it?

Robert: Yes, you're right. This particular item of plant is a well approved compact unit, which has a climbing mixer-cum-skip using a two-zone counterflow mixing system. The replaceable tiles lining the drum and the mixing paddles are of high quality wear-resistant steel alloy.

Duncan: Switching to something on a much greater scale, I'd like to mention some interesting plant, which I once saw, for producing mass concrete to cast a large dam across a river. Casting my mind back, I can vividly recall the wide conveyor belt used to mix, as well as to transport the various grades of aggregate and the cement. Furthermore, the conveyor belt was cleverly designed to incorporate weighing machines situated directly beneath the moving belt. Thus it was possible to weigh the different aggregates, in the required proportions, both automatically and continuously.

Robert: Did this system prove successful?

Duncan: Oh yes! It was a great success, especially with dry mix concrete, because of the inclusion of some suitable shovel-like agitators arranged just above the belt surface.

Robert: Of course, there is much scope in concrete plant for an ingenious engineer to utilise such ideas and improvise a little as necessary. I would say, however, that whenever possible, it is generally better to make full use of recommended specially manufactured plant. In this field, there are already many highly developed forms of plant to suit nearly every requirement, both quantity and qualitywise. Before I forget, I really must say a little about the "Johnston Tower", which is, as you know, a notable self-contained concrete weighing and mixing tower. One of the largest types has 14 separate compartments of about 700 m^3 capacity for various aggregates and cement, and is capable of producing 100 m^3 per hour in batches up to 5 m^3.

Duncan: Of course, this is a very large concrete plant – the sort of plant, in fact, which is ideal for feeding a never-ending fleet of ready-mix concrete lorries serving a host of building sites.

Robert: Something else has just crossed my mind, which is of some interest, I'm sure. I'm thinking of those lightweight concretes, using plastic as aggregates, which have been developed to resist frost, amongst other purposes.

Duncan: I understand this economic technique has been employed with success in road construction in Germany and Finland.

Robert: This is true, and to round off this episode, I'd like your comments on a new concept in lightweight concrete known as Wheatcrete. Firstly, I should perhaps explain what Wheatcrete is – that is if the name itself hasn't already given the game away.

Duncan: Well Robert, the name does indeed give the broad idea that wheat is the main aggregate, but maybe you could give a little more detail.

Robert: This is a rather revolutionary Canadian development, which utilises expanded, or puffed wheat as a substitute for other lightweight aggregates. The wheat kernels are dampened, expanded and then carbonized, or burnt to stabilise them and make them biologically inert.

Duncan: My foremost comment is that it is considerably cheaper than any other lightweight aggregate. Secondly, the fact that it is a replenishable raw material, gives it enormous advantages. Wheat when ruined as a cereal by bad weather can now be put to good use in construction, and this in itself makes economic sense.

Robert: I can see you are enthusiastic about its prospects.

Duncan: Yes I am, but of course the proof of the pudding is in the eating, as with so many new ideas, isn't it?

Vocabulary	Wörterverzeichnis	Glossaire
tool	Werkzeug	outil
compacting tool	Verdichtungsgerät	engin de compactage
vibrating poker	Rüttelflasche	tube pervibrateur
to set	abbinden	faire prise
to cure	nachbehandeln	traiter après prise
concrete sample	Betonprobe	éprouvette de béton
bending tension strength	Biegezugfestigkeit	résistance à la traction-flexion
cube test	Würfeldruckprobe	essai de résistance sur cube
as per	laut, gemäß	suivant
breaking load	Bruchlast	charge de rupture
permissible stress, safe stress	zulässige Spannung	contrainte admissible
mould	Form	moule
dynamic ball impact test	Kugelschlagprüfung	essai à bille
subsidy	staatlicher Zuschuß, Subvention	subvention
credit facilities	Kredit zu günstigen Bedingungen	crédit à conditions favorables
interest rate	Zinshöhe, Zinsfuß	taux d'intérêt
bonus	Prämie	prime, bonus
laid off	arbeitslos	sans travail
flash set	blitzschnelles Abbinden	prise rapide
antifreeze admixture	Anti-Frostmittel	produit antigel
cold spell, (cold period)	Kaltwetterperiode	temps froid
freezing point	Gefrierpunkt	point de congélation
rust corrosion	Verrosten	rouillure
frost protection agent	Frostschutzmittel	antigel
efflorescence	Ausblühen	efflorescence
sodium	Natrium, Natron	sodium
out of this world, (extraordinary, fantastic)	an Wunder grenzend	extraordinaire, miraculeux
to blend	mischen	mélanger
to pass the buck, to transfer the responsibility	die Verantwortung zuschieben	reporter la responsabilité
to get it over, (to make it understood)	es verständlich machen	rendre compréhensible
deformability, stiffness	Steife, Steifigkeit	déformabilité
to disintegrate	zerfallen, auflösen	décomposer
to pour, to place	gießen, schütten	verser
unit	Einheit, (Einzel-)Teil	unité, organe

compacting factor	Verdichtungsfaktor	degré de compactage
slump test	Ausbreitversuch	essai d'affaissement
flow table	Ausbreittisch	table à secousse
conical mould	Kegeltrichter	tronc de cône
to go along with, (to agree with)	damit einverstanden sein	être du même avis
ultra fine material	Feinstkorn, Staubkorn	grain très fin
lining	Auskleidung	revêtement
outset, (beginning)	Beginn, Anfang	début
aggregate	Zuschlagstoffe	granulat
surface tension	Oberflächenspannung	tension superficielle
water content	Wassergehalt	teneur en eau
to chip in, (to interrupt)	einwerfen, unterbrechen	interrompre
vibrator	Vibrator, Rüttler	vibrateur
to weld	schweißen	souder
narrow bend	enger Bogen	coude étroit
to clog	verstopfen	colmater
to squeeze	(zusammen)drücken, (aus)quetschen	comprimer
piston	Kolben	piston
hose, tube	Schlauch	tuyeau flexible
blockage	Blockierung, Sperre	blocage
plant	Betriebsanlage, Maschineneinrichtung	équipement
skip	Aufzugskübel, Beschickungskasten	benne de chargement
counterflow	Gegenstrom	contre-courant
tile	(Auskleidungs-)Platte	carreau
drum	Trommel	tambour
mixing paddle	Mischschaufel	palette de malaxeur
steel alloy	Stahllegierung	alliage d'acier
mass concrete	Massenbeton	béton de masse
dam	Talsperre	grand barrage
conveyor belt	Förderband	tapis roulant
weighing machine	Wiegeeinrichtung	appareil de pesée
dry mixing	Trockenmischung	mélange sec
shovel-like	schaufelähnlich	semblable à une pelle
agitator	Rührwerk, Rühreinrichtung	agitateur
compartment	Abteilung, Zelle	compartiment
building site	Baustelle	chantier
to cross one's mind, (to be thought of)	in den Sinn kommen	avoir une idée
plastics	Kunststoff	materiau plastique
road construction	Straßenbau	construction routiere
to round off, (to end)	abschließend sagen	pour finir
to give the game away, (to reveal something)	die Sache verraten	aller raconter
expanded wheat, puffed wheat	Blähweizen	blé expansé
cereal	Getreide	céréale

4. STRUCTURAL STEELWORK

It would be difficult to visualise the construction industry without steel, which is often the principal material providing the necessary stability in modern structures. Even its foremost competitor, reinforced concrete, relies very much on steel, in the form of bars, for its tensile strength. Notable examples, over the years, of structural steelwork forming the basic framework are self-evident in bridges, towers, pylons, and to a lesser extent, in factories and warehouses etc. In house building too, steel is playing an ever increasing role as the main alternative to timber in floor beams, roof trusses etc., not to mention windows, doors, staircases, balustrading and the like.

Steel, as opposed to any other single material, does lend itself to the prefabrication of really large structural units in the factory or workshop. These units, such as portal frames, compound girders, lattice beams etc., can be easily and quickly connected to others on site, and their dismantling or replacement, where necessary, usually presents few problems. All steel members can be properly designed according to the direction of the forces, which are often calculated by modern methods, or even determined with the aid of model static techniques. These techniques are carried out by subjecting transparent plastic miniature reproductions of parts of the structure to scaled down forces in relation to the real forces of nature. Through using such techniques, good steel structures will never be overweighted. On the contrary, with the improved quality of steel, the tendency is for modern structures to become more slender and delicate in appearance, rather like filigree work in a way.

Since steel is a dynamic factor in today's construction industry, it is not only necessary for further improvements in its strength, but also in its degree of resistance against corrosion. In recent years, however, considerable progress has been made in both respects, in that technologists have provided a form of steel, namely stainless steel, which possesses all the best characteristics of ordinary steel, and yet does not deteriorate through rust when exposed to the elements. Stainless steel, if it could be mass produced, might well become available to the industry at an economically realistic price. Bridges, for example, constructed of stainless steel would need very little maintenance, as is only the case nowadays when they are constructed of stonework, concrete or aluminium alloy.

In order to make steel construction as economical as possible, the fabrication processes, the shapes of the profiles and the connecting methods have been improving steadily for some time. Three outstanding and well established achievements in steel production have been the seamless tube process of Mannesman, the Euronorm profiles and arc-welding. When Thomas Edison was asked as long ago as 1893 at the World Exhibition in Chicago, what he considered to be the most significant invention of the time, he replied without hesitation, "the Mannesman seamless tube". Provided the level of research is maintained in the future, a very high standard of steel will doubtless be attained over a wide field of the industry.

Steel Production

Steel is produced from pig iron by smelting iron ore, which is rarely found pure in its natural state. The smelting process entails heating the iron ore in a blast furnace with coke, which results from intensely heated coal, and limestone, thus reducing it to pig iron. These huge blast furnaces are capable of operating day and night non-stop, often for periods exceeding 10 years. The oxygen required for the burning process is supplied by hot air blown into the lower part of the furnace. The coke serves two purposes, one to provide a burning fuel and another to reduce the iron ore to pig iron. The liquid slag, which is also produced in the furnace, consists mainly of the non-metallic constituents of the ore, like silica, combined with the lime from the limestone. On solidifying, this by-product is utilised for a number of purposes, such as filling under roads etc., the basic material for producing paving blocks, and in powder form, one of the ingredients of Portland blast furnace cement.

The pig iron from the blast furnace is brittle, mainly because of its high carbon content, which is between 3 % and 4 %. To produce steel, which is more rigid in character, this content must be reduced to below

1.7 %. Furthermore, other disagreeable ingredients, like sulphur, phosphorous and silicon must also be eliminated. These requirements can be achieved to varying degrees by a number of different methods, three of which are as follows:–

(a) the Bessemer or Thomas process, using a converter,
(b) the open hearth process, or
(c) the electric furnace process.

The above processes (a) and (b) are employed to mass produce standard quality steel, whereas with process (c) it is even possible to produce special high quality steel, containing such alloy additives as manganese, chromium, nickel, tungsten, molybdenum, vanadium, copper and aluminium. By using process (b) or (c), scrap iron can be reused in the production of steel.

Steel Sections

Structural steelwork basically consists of standard sections and plates. The principal sections, usually manufactured from hot rolled mild steel, are universal beams, rolled steel joists (r.s.js.), universal columns with parallel flanges, channels, equal and unequal angles and T- and Z-bars. Besides these, hollow circular sections are also very popular and economical, because they are ideal for any load support. Nature itself has proved the suitability of this particular section in its use for the various stems and stalks of flowers and crops. Hollow square and rectangular sections are also economical, particularly because the connections between them are easily made. Many of the sections mentioned are also available cold formed to similar shapes.

All steel sections are available in varying grades according to the properties of the steel and the subsequent treatment. The British Standards (B. S.), for example, have three main grades, namely 43, 50 and 55, and the permissible working stresses are given for each. Fig. 10 shows extracts from the handbook of the British Constructional Steelwork Association Ltd., in which all the necessary information for the structural steel designer and erector is given. The terms flange, web, slope and root radius, which are common to most sections, are referred to in this figure.

Connections

(a) Rivets

The connecting together of the members of a steel structure can be carried out in three quite different ways, namely riveting, which is the least common nowadays, bolting and welding, Most of the rivets used for structural steelwork are hot driven shop or field rivets, and they simply consist of a short length of ductile and malleable steel with an almost semi-spherical cup or snap head. Alternatively, the heads are sometimes knobbled or countersunk. As the rivet in its plastic state is being driven into place with a riveting hammer, a second head is formed by the blows and the rivet usually swells and completely fills the hole. During the hammering process, a bucking tool, or dolly, as it is also known, is placed under the rivet head at the back of the connection, in order to hold the rivet steady and serve as an anvil.

After being hammered into place, the rivet cools and tends to shorten, and this provides compression, or a clamping action, between the steel surfaces in contact with each other. In other words, a frictional resistance is set up to prevent the movement of one member against the other. Loose or otherwise defective rivets can be detected by a ringing sound from the surrounding metal when they are tapped with a light hammer. Prior to riveting, the members must be temporarily bolted in positions.

(b) Bolts

Although bolting is more expensive than riveting, it has, nevertheless, an ever growing use in most forms of steel construction, and this applies not only to site connections. A bolt consists of a metal pin with a head at one end and a male threaded shank at the other to receive a female threaded nut. To connect one member to another, bolts are pushed through aligned holes in each and nuts are threaded on and tightened up with a spanner, until the connected members are held firmly together. Bolts should be tightened to such an extent that at least one clear thread projects beyond the nut.

BRITISH CONSTRUCTIONAL STEELWORK ASSOCIATION LTD 85 R

JOISTS

DIMENSIONS FOR DETAILING

Nominal Size	Mass per metre	Depth of Section D	Root Radius r_1	Toe Radius r_2	Flange: Width B	Flange: Thickness T	Flange: Inside Slope	Web: Thickness t	Web: Depth d	Web: $\frac{t}{2}$ + 2mm	Notch: N	Notch: n
mm	kg	mm	mm	mm	mm	mm	degrees	mm	mm	mm	mm	mm
203 × 102	25.33	203.2	9.4	3.2	101.6	10.4	5	5.8	161.0	4.9	55	25
178 × 102	21.54	177.8	9.4	3.2	101.6	9.0	5	5.3	138.2	4.7	55	25
152 × 89	17.09	152.4	7.9	2.4	88.9	8.3	5	4.9	117.9	4.5	45	20
127 × 76	13.36	127.0	7.9	2.4	76.2	7.6	5	4.5	94.2	4.3	40	20
102 × 64	9.65	101.6	6.9	2.4	63.5	6.6	5	4.1	73.2	4.1	35	20
76 × 51	6.67	76.2	6.9	2.4	50.8	5.6	5	3.8	50.3	3.9	30	15

CHANNELS

DIMENSIONS FOR DETAILING

Nominal Size	Mass per metre	Depth of Section D	Root Radius r_1	Toe Radius r_2	Flange: Width B	Flange: Thickness T	Flange: Inside Slope	Web: Thickness t	Web: Depth d	Web: t + 2mm	Notch: N	Notch: n
mm	kg	mm	mm	mm	mm	mm	degrees	mm	mm	mm	mm	mm
432 × 102	65.54	431.8	15.2	4.8	101.6	16.8	5	12.2	362.5	14.2	105	40
381 × 102	55.10	381.0	15.2	4.8	101.6	16.3	5	10.4	312.4	12.4	105	40
305 × 102	46.18	304.8	15.2	4.8	101.6	14.8	5	10.2	239.3	12.2	105	35
305 × 89	41.69	304.8	13.7	3.2	88.9	13.7	5	10.2	245.4	12.2	90	35
254 × 89	35.74	254.0	13.7	3.2	88.9	13.6	5	9.1	194.8	11.1	90	35
254 × 76	28.29	254.0	12.2	3.2	76.2	10.9	5	8.1	203.7	10.1	80	30
229 × 89	32.76	228.6	13.7	3.2	88.9	13.3	5	8.6	169.9	10.6	90	35
229 × 76	26.06	228.6	12.2	3.2	76.2	11.2	5	7.6	178.1	9.6	80	30
203 × 89	29.78	203.2	13.7	3.2	88.9	12.9	5	8.1	145.3	10.1	90	35
203 × 76	23.82	203.2	12.2	3.2	76.2	11.2	5	7.1	152.4	9.1	80	30
178 × 89	26.81	177.8	13.7	3.2	88.9	12.3	5	7.6	120.9	9.6	90	35
178 × 76	20.84	177.8	12.2	3.2	76.2	10.3	5	6.6	128.8	8.6	80	30
152 × 89	23.84	152.4	13.7	3.2	88.9	11.6	5	7.1	97.0	9.1	30	30
152 × 76	17.88	152.4	12.2	2.4	76.2	9.0	5	6.4	105.9	8.4	30	30
127 × 64	14.90	127.0	10.7	2.4	63.5	9.2	5	6.4	84.1	8.4	65	25
102 × 51	10.42	101.6	9.1	2.4	50.8	7.6	5	6.1	65.8	8.1	55	20
76 × 38	6.70	76.2	7.6	2.4	38.1	6.8	5	5.1	45.7	7.1	40	20

The dimension **N** in the case of joists is equal to half the flange width, i.e. B/2 and in the case of channels is equal to the flange width **B** to the nearest 5mm above.

BRITISH CONSTRUCTIONAL STEELWORK ASSOCIATION LTD 269 R

BASED ON BS 449 1969

JOISTS

DIMENSIONS AND PROPERTIES
GRADE 43 STEEL

Actual Size D×B mm	Critical Span Le metres	Area of section cm^2	Moment of Inertia: Axis x–x cm^4	Moment of Inertia: Axis y–y cm^4	Radius of Gyration: Axis y–y cm	Elastic Modulus: Axis x–x cm^3	Elastic Modulus: Axis y–y cm^3	Ratio D/T
203.2 × 101.6	2.259	32.26	2294	162.6	2.25	225.8	32.02	19.5
177.8 × 101.6	2.257	27.44	1519	139.2	2.25	170.9	27.41	19.8
152.4 × 88.9	2.053	21.77	881.1	85.98	1.99	115.6	19.34	18.4
127.0 × 76.2	1.856	17.02	475.9	50.18	1.72	74.94	13.17	16.7
101.6 × 63.5	1.621	12.29	217.6	25.30	1.43	42.84	7.97	15.4
76.2 × 50.8	1.396	8.49	82.58	11.11	1.14	21.67	4.37	13.6

BASED ON BS 449 1969

CHANNELS

DIMENSIONS AND PROPERTIES
GRADE 43 STEEL

Actual Size D×B mm	Critical Span Le metres	Area of section cm^2	Moment of Inertia: Axis x–x cm^4	Moment of Inertia: Axis y–y cm^4	Radius of Gyration: Axis y–y cm	Elastic Modulus: Axis x–x cm^3	Elastic Modulus: Axis y–y cm^3	Ratio D/T	Dimension p cm
431.8 × 101.6	2.547	83.49	21399	628.6	2.74	991.1	80.15	25.7	2.32
381.0 × 101.6	2.718	70.19	14894	579.8	2.87	781.8	75.87	23.4	2.52
304.8 × 101.6	2.874	58.83	8214	499.5	2.91	539.0	66.60	20.6	2.66
304.8 × 88.9	2.377	53.11	7061	325.4	2.48	463.3	48.49	22.2	2.18
254.0 × 88.9	2.643	45.52	4448	302.4	2.58	350.2	46.71	18.7	2.42
254.0 × 76.2	2.011	36.03	3367	162.6	2.12	265.1	28.22	23.3	1.86
228.6 × 88.9	2.785	41.73	3387	285.0	2.61	296.4	44.82	17.2	2.53
228.6 × 76.2	2.163	33.20	2610	158.7	2.19	228.3	28.22	20.4	2.00
203.2 × 88.9	2.945	37.94	2491	264.4	2.64	245.2	42.34	15.8	2.65
203.2 × 76.2	2.320	30.34	1950	151.4	2.23	192.0	27.59	18.1	2.13
177.8 × 88.9	3.117	34.15	1753	241.0	2.66	197.2	39.29	14.5	2.76
177.8 × 76.2	2.389	26.54	1337	134.0	2.25	150.4	24.73	17.3	2.20
152.4 × 88.9	3.326	30.36	1166	215.1	2.66	153.0	35.70	13.1	2.86
152.4 × 76.2	2.400	22.77	851.6	113.8	2.24	111.8	21.05	16.9	2.21
127.0 × 63.5	2.275	18.98	482.6	67.24	1.88	75.99	15.25	13.8	1.94
101.6 × 50.8	1.828	13.28	207.7	29.10	1.48	40.89	8.16	13.4	1.51
76.2 × 38.1	1.573	8.53	74.14	10.66	1.12	19.46	4.07	11.2	1.19

Tabular loads marked thus + are based on the maximum shear value of the web and are less than the permissible flexural load.
Tabular loads printed in italic type are within the web buckling capacity of the unstiffened web and produce a total deflection not exceeding 1/360th of the span.
Tabular loads printed in ordinary type should be checked for deflection. see page 110.
For explanation of tables see notes commencing page 102.
1 kilonewton may be taken as 0.102 metric tonne (megagramme) force, but see page 102.

Fig. 10: Pages from Steelwork Handbook

(Reproduced by permission. British Constructional Steelwork Association Ltd.)

Structural bolts are generally classified according to the following:–

(a) whether they are of ordinary structural or high tensile strength steel.
(b) whether they are regular or heavy duty.
(c) whether they are unfinished or with a turned shank.
(d) whether they are hexagonally, or to a lesser extent, square headed, and
(e) the pitch and fineness of thread.

Ordinary structural bolts are made of mild steel, whereas high tensile strength bolts are made of quenched and tempered steel having a greater yield point than mild steel. Unless there are special requirements, bolts normally have washers of 5 mm thickness and well forged hexagonal heads and nuts. Tapered washers are used to provide a flat bearing on sloping surfaces, as on flanges for example. It is important to employ bolting where the structure may have to be dismantled, where post-tensioning is necessary, or where vibration, as occurs with riveting, is not permitted, either because of the nature of the structure or because of the disturbance factor.

Black bolts are normally quite adequate for site connections, but close tolerance turned bolts, known also as bright bolts, should be used for structures where even the slightest displacement is not permitted. The main difference between black and bright bolts is that the plain shank of the latter has been specially machined to give it a larger diameter than the threaded part. In other words, only a very small tolerance exists between the shank and the hole. It is essential that all bolt holes should be accurately drilled and any burring filed away before bolting commences.

Normally, the forces and moments in bolted, as in riveted connections too, are carried over from one member to the other in the form of shearing forces and by the pressure of the bolt shanks against the surfaces around the holes. In practice, the bolts would also be strained by bending. Through the friction of one steel surface pressing against another, the pressure of the two connected parts gives added assistance in conveying these forces. As with normal bolts, this is never taken into consideration in the calculation of forces.

By using high strength friction grip bolts, the connections are mainly made through the friction between the steel surfaces caused by the clamping action of the bolts. These bolts are placed in the holes of slightly larger diameter to permit a push fit. An adequate friction force can only be attained if the surfaces in contact are prepared beforehand by either flame or blast cleaning. To ascertain this calculated friction force, a defined torque, which is obtainable from standard tables and is dependent on the acquired connection force and the bolt diameter, must be applied to each bolt. The required torque can be guaranteed by using special tools with adjustable torque rates, like a calibrated torque limiting spanner or a pneumatic impact wrench set to a specified torque valve.

(c) Welding

The majority of shop connections are welded nowadays, and the most significant and effective method is electric arc welding, although oxyacetylene welding is still employed. The arc welding process can be carried out both manually, using coated electrodes, or through a suitable automatic means. Normally, a direct electric current (d. c.) of between 50 and 60 volts is used for welding in steel construction. This is either supplied by a d. c. generator usully driven by a 3-phase alternating electric current (a. c.) motor, or a transformer and rectifier. The positive terminal must be connected to the base metal and the negative one to the electrode.

At the start of the process, the electrode must be in direct contact with the base metal, also termed the parent metal or workpiece. Afterwards, it should be held a short distance away, the purpose of which is to strike an electric arc. Hot liquid electrode steel at a temperature of up to 4 000° Celsius then passes along this arc from the tip of the electrode to the base metal. In this way, for example, it is possible to fill a small gap between two pieces of steel with hot molton material, and at the same time fuse this material with the melted edges of the steel. Because of the very high temperature, this process can be executed very quickly, and it is therefore unnecessary to heat up a large area around the welded joint. For this reason, it is economical, and above all it avoids any possible damage through intense heat to adjacent members. Gas welding, which creates such intense heat around the joint, can often impair the quality of the steel.

There are several different types of electrode, depending on the grade of steel to be welded, and the kind of coating on an electrode wire or rod is most important, expecially with so many so-called high efficiency grades on the market. Provided high efficiency electrodes are carefully selected, there is little doubt that the working time can often be cut by half for downhand and production welding. They also provide better welding properties than are possible with ordinary electrodes.

There are two basic types of weld, the butt and the fillet weld, and Fig. 11 shows sections through joints prepared to receive the most common butt welds. The square butt weld, which is employed to make the joint as shown at (a), is only used where the thickness of the base metal does not exceed 5 mm. The use of a single V butt weld, for the joint at (b), is uneconomical where the base metal does not exceed 15 mm, and therefore a double V butt weld, for the joint at (c), is more suitable in most cases. When employing the double V butt weld, the process must be performed alternately from both sides. Apart from the small sealing run at the root, the single U butt weld, for the joint at (d), can be produced from one side only without losing the very advantage of employing a double V butt weld. This important advantage is that no shrinkage of the weld occurs, because of the even distribution of temperature, and therefore the joint does not become cranked. For base metals exceeding 30 mm in thickness, it is advisable to use double U butt welds or symetrical UV butt welds, mainly because electrode material is saved.

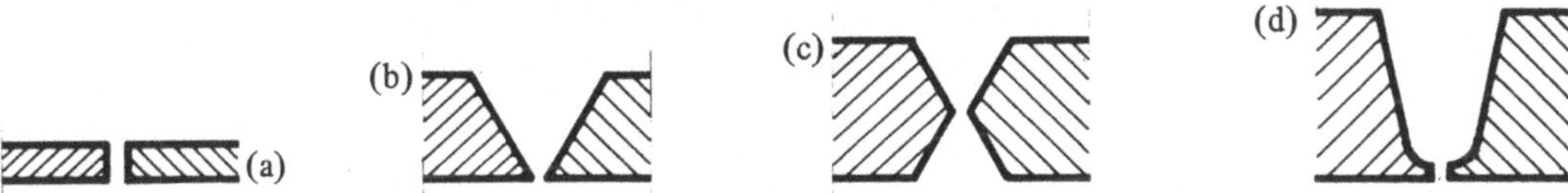

Fig. 11: Sections through Joints to receive Butt Welds

To weld a vertical member to a horizontal one, or vice-versa, it is possible to use either a double fillet weld or a K weld, although in some cases a single fillet weld is sufficient for this purpose. The visible side of a fillet weld, known as the face, can be finished convex, concave or flat, the last being the most common method, because it gives the same strength, using less electrode, as the other two. If a concave fillet weld is deeply hollowed out, it is the best form for the flow of forces, but it is the most difficult to produce. In the design of butt welds, the thickness of the throat is equal to the thickness of the thinnest base metal to which it is applied. For fillet welds, on the other hand, only the base width need equal this base metal thickness.

All steel structures must be accurately designed and properly erected in order to best serve their purpose. A well designed and erected welded structure, for example, can often give as much as a 30 % saving over a similar riveted construction. One of the less obvious reasons for this is because the smooth surface area afforded by a welded framework makes the painting a far easier and quicker process, thus reducing its cost. Any contractor who undertakes welding must be fully authorised to do so. He is obliged to test all main welds, and any others required by the engineer, with x-ray or gamma-ray apparatus. He must also provide and instal all the necessary instruments for measuring the deflection and stresses as directed by the engineer. Generally speaking, theory is important in the design and erection of a welded steel structure, but the optimum results depend very much on the skill of the welder. Even if the welding is carried out fully or semi-automatically by machine, the operator has the ultimate control over the precision of the work.

Structures

The steel skeleton of a modern multi-storey building consists of a rigid stable construction capable of carrying all dead and live loads, including those resulting from wind and snow, and transmitting them to the foundations. Even though mild steel is the most commonly employed, high tensile steel is sometimes more economical, because its higher yield point permits greater working stresses and thus it reduces the weight of the framework. However, deflection criteria often limit this significant advantage of high tensile steel. A steel framework, rather in the form of a cage, constitutes a hyperstatic system, and its calculation and design used to be quite a laborious task. Nowadays, with the use of computers, this work has been considerably reduced.

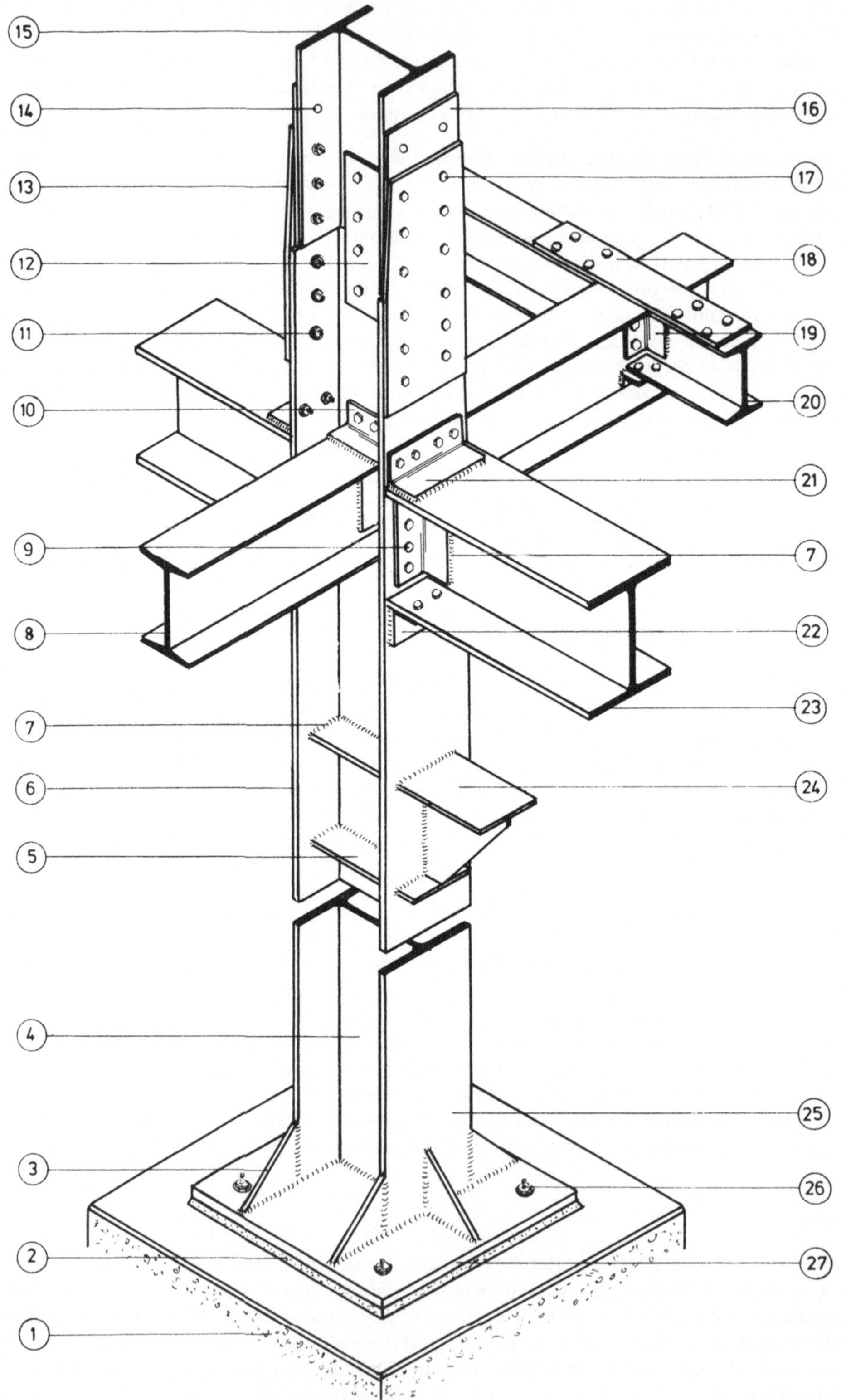

Fig. 12: Steel Stanchion with Girders and Joists

	stanchion, (structural) steel column	Stahlstütze	poteau en acier
	girder	Träger	poutre
	joist	Querbalken	poutre transversale
1	**concrete stanchion or column base**	Betoneinzelfundament für Stütze	fondation individuelle pour poteau
2	**cement and sand grouting**	Zementmörtellage, Feinbetonschicht	couche de mortier de ciment
3	**triangular stiffener plate**	Aussteifungsdreieck	triangle de renforcement
4	**web**	Steg	âme
5	**horizontal stiffener plate**	waagerechtes Aussteifungsblech	plaque horizontale de renforcement
6	**universal column with parallel flanges**	IPB-Träger, Breitflanschträger	poutre à ailes larges type IPB
7	**fillet weld, fillet welding**	Kehl(schweiß)naht	cordon de soudure
8	**rolled steel joist, (r.s.j.)**	I-Träger, Normalprofil	poutre à profil normal type I
9	**black bolt for site connection**	rohe Schraube für Baustellenan-anschluß	boulon pour assemblage sur le chantier
10	**top flange angle cleat**	Anschlußwinkel für oberen Flansch	équerre de raccordement supérieur
11	**nut and washer of site bolt**	Mutter und Unterlegscheibe für Montageschraube	écrou et rondelle d'assemblage
12	**web fishplate**	Steg(stoß)lasche	éclisse d'âme
13	**flange fishplate**	Flansch(stoß)lasche	éclisse d'aile
14	**shop rivet**	Werkstattniet	rivet venu d'usine
15	**smaller section of No. 6 above**	kleineres Profil von Nr. 6,	profil réduit du n° 6
16	**steel packing plate**	Futterblech	tôle réductrice
17	**turned bolt, tight fitting bolt, bright bolt**	Paßschraube	boulon
18	**tension plate**	Zuglasche	plaque de renforcement
19	**web angle cleat, connection angle to web**	Anschlußwinkel an Steg	équerre de raccordement sur âme
20	**floor joist**	Deckenträger	poutre de plancher
21	**top flange angle cleat**	Anschlußwinkel an oberen Flansch	équerre de raccordement supérieur
22	**seating angle cleat**	Anschluß- und Auflagewinkel	équerre de raccordement et d'appui
23	**broad-flanged girder**	Breitflanschträger	poutre à ailes larges
24	**steel bracket cut to size from broad-flanged section and welded directly to stanchion**	Breitflanschträgerabschnitt als Stahlkonsole direkt an Stahlstütze geschweißt	section de poutre à ailes larges servant de console et soudée directement sur le poteau
25	**flange**	Flansch	aile
26	**anchor bolt, holding down bolt**	Ankerschraube	boulon de fixation
27	**steel base plate**	Fußstahlplatte	plaque de base

If there is some doubt about the anticipated loads and their distribution, because of the indeterminable degree of restraint of the members at the joints, it is sometimes advisable to use standard tables in the design of a steel framework. This also applies in cases where the soil resistance is particularly uncertain. Where such tables are used in preference to normal calculation and design, it is important, however, that they are based on well founded calculations and have been proved absolutely satisfactory in past practice.

These days, it is not considered economical to have a framework made up of too many different sized members, where this entails much labour with complicated connections and special packing pieces etc. A simple construction with as few variations in section sizes as possible is preferable. Such a construction is almost entirely welded together with the exception of the joints of the larger shopwelded units. These are normally bolted together and sometimes site welded afterwards in addition. A direct welded connection of one member to another without using cleats or plates is seldom carried out on site. To make a connection of this kind on site would necessitate accurate cutting to length and would complicate the erection process.

It is therefore more common and practical to use flange and web cleats to overcome these problems. The degree of stiffness using such cleats depends on the thickness and length of the cleat leg. To improve the rigidity at the critical joints, gussets and stiffeners are often also welded on.

Vocabulary	Wörterverzeichnis	Glossaire
structural steelwork	Stahlbau	construction en acier
reinforced concrete	Stahlbeton	béton armé
bar	Stab	barre
tensile strength	Zugfestigkeit	résistance à la traction
basic framework	Grundskelett, Gerüstskelett	ossature
pylon	Pylon, Kabelstütze	pylône
factory	Fabrik	usine
warehouse	Lagerhaus	entrepôt
floor beam	Deckenbalken	poutre de plancher
roof truss	(mehrgliedriger) Dachbinder	ferme de toit
staircase	Treppe	escalier
balustrading	Brüstungsgeländer	balustrade
prefabrication	Vorfertigen	préfabrication
structural unit	Bauteil	élément
workshop	Werkstatt	atclier
portal frame	Portalrahmen	portique
compound girder	zusammengesetzter Träger	poutre composée
lattice beam	Fachwerkträger	poutre en treillis
dismantling	Demontage, Auseinanderbau(en)	démontage
scaled down	im (Modell-)Maßstab verringert	réduit
stainless steel	korrosionsbeständiger Stahl	acier inox
rust	Rost	rouille
stonework, masonry	Natursteinmauerwerk	maçonnerie en pierres
aluminium alloy	Aluminiumlegierung	alliage d'aluminium
seamless tube	nahtloses Rohr	tube sans soudure
arc-welding	Lichtbogenschweißen	soudage à l'arc
pig iron	Roheisen	fer brut
iron ore	Eisenerz	minerai de fer
blast furnace	Hochofen	haut-fourneau
limestone	Kalkstein	calcaire
oxygen	Sauerstoff	oxygène
burning fuel	Brennstoff	combustible
liquid slag	flüssige Schlacke	laitier liquide
non-metallic constituent	nichtmetallischer Bestandteil	composant non-métallique
silica	Siliziumdioxid, Quarz	quartz, dioxyde
lime	Kalk	chaux
to solidify	erstarren	solidifier
by-product	Nebenprodukt	sous-produit
paving block	Pflasterstein	pavé
blast furnace cement	Hochofenzement	ciment de haut-fourneau
carbon content	Kohlenstoffgehalt	teneur en carbone
sulphur	Schwefel	soufre
silicon	Silizium	silicium
open hearth process	Siemens-Martinverfahren	procédé Siemens-Martin
electric furnace	Elektroofen	four électrique
standard quality	Handelsgüte	qualité normale
special high quality steel	hochwertiger Stahl	acier à haute résistance
additive	Zusatz	additif
manganese	Mangan	manganese
tungsten	Wolfram	wolfram

scrap iron	Eisenschrott	ferraille
steel section	Stahlprofil	acier profilé
standard section	Normenprofil	profilé normalisé
plate	Grobblech	grosse tôle
sheet	Feinblech	tôle fine
mild steel	Flußstahl	acier doux
hot rolled mild steel	Walzstahl	acier laminé
rolled steel joist, (r.s.j.)	I-Träger, Normalprofil mit etwa 80 bis 200 mm Trägerhöhe	poutre en double T d'une hauteur de 80 à 200 mm
universal beam	I-Träger mit Trägerhöhen von etwa 200 mm an aufwärts	poutre en double T d'une hauteur minimum de 200 mm
universal column with parallel flanges	IPB-Träger, Breitflanschträger	poutre à ailes larges
parallel flanges	parallele Flansche	membrures paralléles
channel	U-Profil	fer en U
equal angle	gleichschenkliges Winkelprofil	cornière à ailes égales
unequal angle	ungleichschenkliges Winkelprofil	cornière à ailes inégales
T-bar	T-Profil	profilé en T
Z-bar	Z-Profil	profilé en Z
hollow circular section	Rohrprofil	profilé creux circulaire
load support	Laststütze, Auflager	support
stalk	Stiel, Stengel, Halm	âme
hollow square section	quadratisches Hohlprofil	profilé carré creux
rectangular	rechteckig	rectangulaire
permissible working stress, allowable stress	zulässige Spannung	contrainte admissible, contrainte autorisée
erector	Aufsteller, Montagefachmann	monteur
flange	Flansch	bride
web	Steg	âme
slope	Neigung	inclinaison
root radius	Ausrundungshalbmesser	rayon de bassage
fastener, connector, connection	Verbindung	liaison, connexion
to rivet	nieten	river
to bolt	schrauben	visser
to weld	schweißen	souder
rivet	Niet	rivet
shop rivet	Werkstattniet	rivet d'atelier
field rivet	Baustellenniet	rivet de chantier
ductile	dehnbar	dilatable
malleable	verformbar	déformable
semi-spherical	halbrund	hémispherique
cup head, snap head	Halbrundkopf	tête demi-ronde
knobbled head	Flachrundkopf	tête en goutte de suif
countersunk head	Senkkopf	tête noyée
riveting hammer	Niethammer	marteau-riveur
bucking tool, dolly	Gegenhalter	tas
anvil	Amboß	enclume
clamping action	Klemmwirkung	action de serrage
frictional resistance	Reibungswiderstand	résistance à la friction
site connecting	Baustellenanschluß	connexion de chantier
bolt	Bolzen	boulon
pin	Stift	goujon
male threaded	mit Außengewinde	avec un filet extérieur
shank	Schaft	fût
female threaded	mit Innengewinde	avec un filet intérieur
nut	Mutter	écrou
aligned holes	ausgerichtete (Bohr-)Löcher	trous alignées

to tighten up	anziehen	serrer
spanner	Schraubenschlüssel	clé à écrous
thread	Gewinde	filet
structural bolt	Stahlbauschraube	boulon
ordinary structural steel	gewöhnlicher oder normaler Baustahl	acier commercial
high tensile strength steel	hochwertiger Stahl	acier de haute résistance
high strength bolt	hochfeste (H.V.) Schraube	vis de haute résistance
unfinished bolt, black bolt	rohe Schraube	vis brute
turned shank	gedrehter Schaft	fût tourné
hexagonally headed	mit Sechskantkopf	avec tête hexagonale
square headed	mit Vierkantkopf	avec tête carrée
pitch	Steigung	pas
to quench	abschrecken	tremper
to temper	anlassen	adoucir
quenched and tempered	vergütet	traité
yield point	Fließgrenze, Streckgrenze	limite d'allongement
washer	Unterlegscheibe	rondelle
forged	geschmiedet	forgé
tapered washer	Keilscheibe	rondelle en forme de coin
to dismantle	demontieren, auseinandernehmen	démonter
post-tensioning	Nachspannen	réglage ultérieur de la précontrainte
close tolerance turned bolt, bright bolt	Paßschraube	boulon ajusté
to machine	maschinell bearbeiten	usiner
to drill	bohren	forer
burring, burr	Grat	barbe
to file away	abfeilen	limer
shearing force	Scherkraft	effort tranchant
to strain	beanspruchen	solliciter
bending	Biegung	flexion
high strength friction grip bolt, high strength bolt	hochfeste Schraube für gleitfreie Verbindung, hochfeste (und vorgespannte) Schraube, H. V.-Schraube	vis de haute résistance
push fit, (slightly tight fit)	genaue Passung	ajustage correct
friction force	Reibungskraft, Reibung	friction
flame cleaning	Reinigen mit Flammstrahl	nettoyage au chalumeau
blast cleaning	Reinigen mit Sandstrahl	sablage
torque	Drehmoment	couple
torque rate, torque value	Drehmomentenwert	taux de couple
calibrated torque limiting spanner, calibrated torque wrench	Drehmomentenschlüssel, Momentenschlüssel mit Anzeige	clef dynamométrique
pneumatic impact wrench set	Druckluft-Schlagschrauber	clé à choc pneumatique
shop connection	Werkstatt-Verbindung	joint d'atelier
oxyacetylene welding, gas welding	Sauerstoff-Azetylen-Schweißen, Gasschweißen, autogenes Schweißen	soudure autogène oxyacétylénique soudage oxyacétylénique
coated electrode	Mantelelektrode	électrode enrobée
direct electric current, (d. c.)	Gleichstrom	courant continu
alternating electric current,(a. c.)	Wechselstrom	courant alternatif
3-phase alternating current	dreiphasiger Wechselstrom, Drehstrom	courant triphasé
transformer	Transformator	transformateur
rectifier	Gleichrichter	redresseur de courant
positive terminal	positive Polklemme	borne positive
base metal, parent metal, workpiece	Werkstück	pièce à travailler

downhand welding	Abwärtsschweißen	soudage du haut vers le bas
production welding	industrielles Schweißen	soudage industriel
butt weld	Stumpfnaht	soudure à rapprochement
fillet weld	Kehlnaht	soudure en gorge
square butt weld	I-Naht	soudure en I
single V butt weld	V-Naht	soudure en V
double V butt weld	X-Naht	soudure en X
sealing run	Kappnaht	soudure d'étanchéité
single U butt weld	U-Naht, Tulpen- oder Kelchnaht	soudure en U
shrinkage	Schrumpfen	retrait
cranked, bent	verbogen, verformt	déformé
double fillet weld	Doppelkehlnaht	soudure en gorge double
K weld	K-Naht	soudure en K
face	Ansichtseite	face apparente
deeply hollowed out	tief(gehend) ausgefüllt	profondement rempli
throat	Nahtdicke	épaisseur de soudure
welded framework	geschweißte Rahmenkonstruktion	bâti soudé
painting	Anstrich	couche de peinture
x-ray	Röntgenstrahlen	rayons X
deflection	Durchbiegung	déflexion
stress	Spannung	contrainte
structural engineer	Bauingenieur	ingénieur civil
welder	Schweißer	soudeur
steel skeleton	Stahlskelett	ossature métallique
multi-storey building	Gebäude mit mehreren Stockwerken	immeuble
dead load	Eigengewicht	poids propre
live load	Verkehrslast	surcharge
foundation	Gründung, Fundament	fondation
hyperstatic system	statisch unbestimmtes System	système hyperstatique
restraint	Einspannung	encastrement
joint	Stoß, Knotenpunkt	joint
soil resistance	Bodenwiderstand, Bodengegendruck	résistance du sol
different sized members	Glieder mit verschiedenen Abmessungen	éléments à dimensions différentes
packing piece	Futter, Ausgleichstück	fourrure
shop-welded unit	werkstattgeschweißtes Teil	élément soudé en atelier
site-welded	baustellengeschweißt	soudé au chantier
cleat, angle cleat, angle bracket	Anschlußwinkel	cornière d'attache
plate, connecting plate	Lasche	éclisse
flange cleat	Flansch-Anschlußwinkel	cornière d'attache de bride
web cleat	Steg-Anschlußwinkel	cornière d'attache de l'âme
stiffness, rigidity	Biegesteifigkeit	rigidite à la flexion
cleat leg	Winkelschenkel	aile de corniere
gusset, gusset plate	Knotenblech	gousset en tôle
stiffener, stiffener plate	Aussteifungsblech	tôle raidisseuse

Dialogue

Duncan: When thinking about structural steelwork, my immediate thoughts usually turn to those large outstanding structures, like the Eiffel Tower in Paris and the many well known long span bridges constructed almost entirely of steel. However, there are numerous less notable achievements in this particular field, which one shouldn't really overlook.

Robert: No doubt, Duncan, you're mainly referring to the hidden steel framework of multi-storey buildings. Once the steel members are encased in concrete, or covered in some other way, the roofing fixed and the infill panels or curtain walling erected between the stanchion and beams, the framework is often soon forgotten. It's really a case of being out of sight, out of mind, as the saying goes. Apart from this very common form of structural steelwork, there are many other forms, to my mind, which are rather taken for granted.

Duncan: Which forms then do you feel are worthy of special mention, Robert?

Robert: Well, to begin with, let's take some of the more common types of tower. Perhaps the best known is the pylon for the distribution of high voltage electricity. The main framing and bracing are usually constructed of hot-dip galvanised mild steel angle sections. Another common tower, which I suppose should strictly be regarded as working plant rather than a structure, is the tower crane. In this case, the construction is generally in the form of mild steel lattice box beams, both for the tower sections and the jib.

Duncan: Then, of course, we have oil well derricks and oil fractionating towers, and if we are to be smack up to date, we should just mention the rocket launching towers of our space age.

Robert: In fact, if one cares to look at the world of steel in an even broader sense, such items as storage bins, bunkers and silos, caissons, gasholders, barges and a whole host of others cannot be ignored.

Duncan: You're quite right, but the construction of a barge, for example, hardly comes into the realms of our particular business, you must admit.

Robert: Perhaps not, but nevertheless I do recommend that every structural engineer should try to visit a large modern ship building yard, where he will certainly learn a great deal to his advantage. The system for connecting steel sections and plates together is similar to that of the construction industry, but the technique does differ in some respects.

Duncan: In such a yard, I understand that the cutting and shaping of really massive plates are often carried out by a fully automatic process. The plates are lifted up one by one and positioned by means of overhead cranes or gantries with electro-magnets. Autogenous flame cutters, which are automatically guided along invisible contour lines, are then employed to cut the plates to size. Are you familiar with this process?

Robert: Yes, I actually saw it when I was once shown around a workshop. I was most interested in the extent of automation generally, but in this particular process the progress has been tremendous. It is now possible to synchronise the flame cutter precisely with the lines on the scale drawing by means of light beam pick-ups and an electronically controlled motor driven device.

Duncan: To fully appreciate this and other modern inovations, one really needs to experience them at first hand. Maybe at this point, you would care to say a little about your own impressions when you witnessed field welding of bridge units somes years ago.

Robert: Well, I remember that occasion quite well. It took place in the Netherlands, where one of a number of steel bridges was being constructed across a waterway. The large units were being welded together on site in filthy weather, and I was most impressed with the efficiency with which a four-man welding team tackled the job. This responsible work was done very accurately, even in those conditions, using a semi-automatic electric arc-welding machine with an astonishingly fast travel speed.

Duncan: Some while ago I also saw a similar machine, but of a smaller type I would imagine. It was being used for fillet welds in the production of crane sections. What fascinated me most of all was the continuous delivery of the wire and flux from the machine. All this, of course, was carried out in the comparative comfort of the workshop. Incidentally, how did they manage the job you spoke of in such foul weather?

Robert: In this particular case, a mobile covering was erected, which simply consisted of scaffolding on wheels covered with a heavy duty plastic sheeting. It's worth mentioning too, in passing, that the welding machine was working with protection gas and an x-ray or gamma-ray controlling device.

Duncan: Despite other very successful forms of bridge construction, steelwork is still called upon from time to time in this vast field. This is especially true in the construction of temporary flyovers in our traffic congested towns. In my view, this is probably the only quick solution to the problem. What do you think?

Robert: To a large extent, I agree. With steel prefabrication and rapid erection techniques, there seems to be no other alternative in providing a flyover quickly with relatively low labour costs and of the required quality. I often chuckle to myself at the expression "temporary flyover", especially when I think of the many years you've had one in your home town.

Duncan: Well, we prefer to call our so-called temporary flyover, semi-permanent.

Robert: Apart from these flyovers, there are, of course, steel bridges which could hardly be built of any other material, like reinforced concrete for example.

Duncan: Sorry to take the words out of your mouth, but you are obviously referring to pipeline bridges and the like. It goes without saying that this type of bridge is nearly always made entirely of steel.

Robert: If you don't mind, I'd like to jump in here and give a short resumé on bridge construction in general.

Duncan: Fire away by all means, but please try not to go back too far into the dim and distant past.

Robert: I promise not to dig up too much history. On the contrary, I'll start by mentioning that in recent years Europe has adopted American ideas more and more in bridge building. This is borne out in the greater degree of standardisation to reduce production costs. Take for example the orthotropic light deck construction, which completely replaces the conventional top girder. Incidentally, this is commonly used for road bridges nowadays.

Duncan: Could you fill our readers in with a few details?

Robert: Sure. This steel decking is normally surfaced with about 5 cm of mastic asphalt, but where the decking must be particularly lightweight, as with transportable bridges or replaceable flyovers, the surfacing often consists of a little as 2 cm of plastic.

Duncan: I shouldn't think that would last 5 minutes.

Robert: It certainly isn't intended to be hard wearing, but there is a considerable risk, in my view, that it would wear away under heavy traffic before one could say "Jack Robinson", to coin an English phrase.

Duncan: Incidentally, I recently read a rather interesting article in the magazine "acier-stahl-steel" about decking. It stated that a decking must be at least 12 mm thick to ensure the necessary rigidity for an asphalt surfacing. It went on to say that the spacing of the stiffening ribs in both directions on a decking section has now generally become wider. A 300 mm spacing, in fact, was given as the most common.

Robert: That's quite a useful bit of information, Duncan. The box profile system is another fairly modern form of construction, and perhaps I could slip in one or two words about it.

Duncan: Please do, Robert.

Robert: The bottom plate of the box section in this system does, in fact, replace the lower chord of the classic beam bridge. By the way, with wide multi-lane bridges, it's possible to have two or more separate box sections rather than the single box with the decking overhanging each side. This is really an alternative construction to the so-called comb form.

Ducan: What do you think of composite construction for bridges? By that I mean the use of reinforced concrete, usually prestressed in both directions, for the decking, whilst the main structure is of steel.

Robert: Well, the improved rigidity achieved by combining the two materials can often reduce the number of wind and sway braces, which are normally necessary. To continue a little on the subject of composite bridge construction, it's a well known fact that there are many good examples, but calculations have shown that this construction is only economical for medium span bridges. For bridges having wider spans, and for suspension bridges and cable bridges with stiffened beams, the decking should only be of steel, because of the weight restriction. Moving to something else on bridges, you have no doubt heard quite a bit about the failure of several unsupported cantilevered construction methods here on the Continent.

Duncan: Yes, indeed I have, but not only on the Continent. To be perfectly honest, I don't remember the reasons why, nor can I recall what happened afterwards.

Robert: Whatever the causes, we have certainly learnt something, as always, from our mistakes, regrettable though they are. In one of the incidents, the Standards used were definitely found to be incorrect, and these had to be put right for any future calculations in this field. As far as I know, the error concerned buckling and bulging.

Duncan: That's a very intricate subject, on which experience, especially in view of these failures, as well as theory can be most helpful.

Robert: Forgive me if I'm embarking on a rather deep topic, but I understand that lateral buckling in a flange could be avoided by restraining the compression flange of the beam. In this way, the lateral deflection in the flange is hindered. The possibility of web buckling through diagonal compression always exists. According to the rule for rolled steel joists, if the value h/t,which is the distance between flanges divided by the web thickness, is less than 70, there is no danger. There is, however, just one proviso to this rule.

Duncan: What is that?

Robert: It is that the direct bearing stress subjected to the root of the web does not exceed a given value, which is provided in the relevant tables.

Duncan: Very interesting, but not being a civil engineer like you, some of this is a little over my head. Now to get down more to my level of thinking, can we talk about such things as modern methods of rust prevention? Have you any detailed information on this?

Robert: Not specific details, but I am all for discussing this topic in general terms, nevertheless. The question of rust, of course, is always the main argument put up against the use of steel. In my opinion, it's grossly exaggerated.

Duncan: I wholeheartily agree. As we say in Britain, "the proof of the pudding is in the eating", and we all have enough proof of the successful use of steel, even where it is constantly in contact with water. Take, for example, its use in the shipbuilding docks.

Robert: There, Duncan, you have hit on a first class example of steel construction in conditions most conducive to rust. One need only think of the mechanical lifts for ships, launching slipways, and ships themselves. All these structures are naturally protected against rust, but look how ancient some of them are – historic relics, in fact. It should also be remembered that the know-how against rust has improved no end since then.

Duncan: That's quite true, Robert. In any case, surface rust which is spotted at the first or even second glance is no real problem, since it can usually be removed by wire brushing and the steel surface can be effectively treated. It's a problem though when it can't be seen at all, such as might be the case with covered reinforcing bars in a foundation which is affected by aggressive ground water.

Robert: Best of all, the steel should be properly protected to prevent rust getting a hold in the first place. There are a number of effective rust protection processes, like phosphating, stove-enamelling and hot-dip galvanising, but these are rather expensive. Steelwork inside a building, for example, can be easily protected for a lifetime by good painting. Even outdoor steelwork nowadays can be successfully protected for a quaranteed 15 years or so, by using special exterior quality paint or a plastic or non-ferrous metallic surfacing.

Duncan: These days, of course, with changing demands, 15 years represents a lifetime for many both temporary and semi-permanent structures.

Robert: Stainless steel, to my mind, provides the real answer where permanency is of the utmost importance, but here we are up against costs. I am sure it would often be preferred to other materials, like glass, galvanised steel, enamelled steel or plastics, for external cladding panels etc., if it were not for this factor. Incidentally, it's interesting to know that the effectiveness of most stainless steel in steel construction is based on there being a thin layer of rust serving as a protective coating, much as patina does on a copper surface.

Duncan: So much for rust. Maybe we should now chat about fire protection measures for steelwork.

Robert: To be frank, I often think that fire hazards in a steel framed construction are not as serious as many people believe.

Duncan: I am sure you are right. Generally speaking, the dangers of fire breaking out in a modern building today are far less acute than a decade or so ago. In the rather conservative methods still being adopted for much house building, there could well remain good reason for concern. However, nearly all materials produced today are at least fire-resistant to some degree, if not refractory.

Robert: Despite all that you have just said, it is still generally required by the Standards that all bare surfaces of structural steelwork in buildings must be insulated against the heat generated by a possible outbreak of fire. This is because intensely heated steel is reduced in strength, as a result, and the beams and columns become deformed under the weight of the construction carried by them.

Duncan: Such insulation can be a costly business, believe you me, but recent developments in this field have produced some acceptable methods at a lower cost. For example, instead of covering steel members with conventional non-inflammable materials, like gypsum plasterboard, asbestos cement sheeting or special quilting, it is now possible to coat the exposed steel surface with a protective asbestos in a plastic form. This material, together with an incorporated cohesive binder, is applied directly to the surface with a gun – rather like the gunnite technique. A thin coating of it provides similar fire-resisting qualities – that is for about 90 minutes – as a 25 mm thickness of plasterboard.

Robert: That really is a stride forward, especially as such a coating could well be developed to contain effective rust prevention qualities too. In fact, taking an overall look at steelwork, in its modern forms, such as stainless steel, plated and plastic coated steel and lightweight cold formed mild steel for sections and seamless tubes, I'm sure it must have a rosy future.

Duncan: You mentioned cold formed sections, which I believe have a lot to offer. The main advantage lies in their cheaper erection costs, due to their lesser weight. They are thus most effective for the wider spans of modern structures, and this is particularly so when they are connected together by spot welding, bolting or screwing with self-tapping screws in lattice trusses. These sections can be made extremely light to handle, because of the low dead to live load ratio. They also have some value in combination with hot rolled sections. Even for axial loads, cold formed columns are suitable in place of the normal broad-flanged girders. Such columns are made up of two channel sections, either welded back to back or the other way around to form a hollow box section. This box section has a very high torsion resistance, and for even higher loads the hollow centre can be filled with concrete, if necessary.

Robert: Your mentioning welding gives me the perfect cue to bring up an odd characteristic of high efficiency electrodes.

Duncan: What's so peculiar about them then?

Robert: Didn't you know that because the coating contains so much iron powder, the weight of the metal recovery is far greater than that of the core wire?

Duncan: No, I didn't realise that, but one lives and learns. It does show that it isn't against the law of nature to gain more from a process than one puts into it. Have you any idea what this gain amounts to?

Robert: As a result of melting down about 100 grams of core wire, we get the astonishing weight of approximately 150 grams of weld metal.

Duncan: The mind boggles at such queer things. Well, to draw yet another absorbing subject to a close, have you any final thoughts about steelwork in general?

Robert: Yes, only to say that steel construction has changed a great deal over the last 50 years or so. For example, since about 1939, very little riveting has been done on bridges. Welding has dominated the scene in this particular branch of the industry.

Duncan: The properties of steel, of course, have improved in leaps and bounds, and I would rate that fact, above all, as the greatest single contribution to progress in the steel world. Even though the three most useful materials, namely timber, steel and reinforced concrete have their individual advantages, the fierce competition between them can prove to be a healthy factor in the future development of all of them.

Robert: It rather sounds as if you have already wound up our discussion, but I'm determined to slip one more point in, come what may.

Duncan: All right, but I think this should be the last one, Robert.

Robert: Fair enough, Duncan. It's simply to say that, in my opinion the gluing together of metal members, and even plain metal surfaces, is coming in gradually. Already, it is being employed in conjunction with arc-welding, and I can see the day when it could well replace it altogether in light metal construction.

Duncan: You may be right, but I think that time is a little way off yet.

Vocabulary	Wörterverzeichnis	Glossaire
long span bridge	weitgespannte Brücke	pont à grande portée
roofing	Bedachung	couverture
infill panel, infilling panel	Ausfachungstafel	panneau de remplissage
curtain walling	Vorhangwand	mur-rideau
high voltage electricity	Hochspannung	haute tension
bracing	Strebe	contrefiche
hot-dip galvanised	feuerverzinkt	zingué au feu
to be smack up to date	ganz aktuell sein	être très actuel
tower crane	Turmdrehkran	grue à tour
lattice box beam	Kastengitterträger	poutre en treillis
jib	Ausleger	flèche
derrick	Derrick, Ladekran	grue derrick
oil well derrick	Bohrturm	derrick
oil fractionating tower	Erdöl-Destillationskolonne	colonne de distillation
rocket launching tower	Raketenabschußrampe	rampe de lancement de fusée
space age	Weltraumzeitalter	ère spatiale
storage bin	Lagerbehälter	réservoir
bunker	Bunker	trémie
caisson	(Druckluft-)Senkkasten	caisson d'air comprimé
gasholder	Gasometer	gazomètre
barge	Lastkahn	péniche
ship building yard	Werft	chantier naval
to cut and shape	formschneiden	couper et former
overhead crane	Laufkran	pont roulant
gantry, gantry crane	Portalkran	portique
autogenous flame cutter	autogener Schneidbrenner	chalumeau-coupeur autogène
contour line	Umrißlinie	ligne de contour
scale drawing	maßstäbliche Zeichnung	dessin à l'échelle
light beam pick-up	lichtelektrischer Abtaster,	cellule photo-électrique
electronically controlled	elektronisch gesteuert	à commande électronique
device	Vorrichtung	dispositif
field welding, site welding	Baustellenschweißung	soudage au chantier
filthy weather, foul weather, lousy weather	Sauwetter, Hundewetter	mauvais temps
semi-automatic	halbautomatisch	semi-automatique
electric arc-welding machine	Lichtbogenschweißgerät	dispositif de soudage à l'arc
flux	Flußmittel	fluidifiant
mobile covering	fahrbare Überdachung	toiture mobile
scaffolding	Gerüst	échafaudage
heavy duty plastic sheeting	schwere Plastikfolie	feuille plastique lourd
protection gas	Schutzgas	atmosphère inerte
temporary flyover	vorläufige Überführung	passage superiéur temporaire
erection	Errichtung, Montage	montage
to chuckle to oneself at something	sich amüsieren über etwas	se moquer de quelque chose
pipeline bridge	Rohr(leitungs)brücke	pont pour canalisations
to fire away (to go ahead)	„loslegen", anfangen	démarrer
to be borne out, to be proven	erwiesen sein	être prouvé
light deck construction	(Brücken-)Tafel in Leichtkonstruktion	plaque orthotrope légère
top girder	oberer Träger	poutre longitudinale
to fill somebody in (to inform someone)	jemanden informieren	informer quelqu'un
hard wearing	verschleißfest	résistant à l'usure

before one could say "Jack Robinson" (very quickly)	„im Handumdrehen"	en un tour de main
to coin a phrase	eine Redewendung anwenden	employer un cliché
asphalt surfacing	Asphalt-Oberschicht, Asphaltdecke	revêtement bitumineux
spacing	Abstand	écartement
stiffening rib	aussteifende Rippe	nervure raidisseuse
lower chord	Untergurt	membrure inférieure
beam bridge	Balkenbrücke	pont-poutres
multi-lane bridge	mehrspurige Brücke	pont à plusieurs voies
composite construction	Verbundkonstruktion	système composite
prestressed	vorgespannt	précontraint
wind brace	Windverbandstrebe	membrure de contreventement
sway brace	Schlingerverband-Strebe	membrure anti-roulis
medium span	mittlere Spannweite	portée moyenne
suspension bridge	Hängebrücke	pont suspendu
cable bridge	Seilbrücke	pont à câbles
stiffened beam	Versteifungsbalken	poutre raidisseuse
unsupported cantilevered construction	Freivorbau-Konstruktion	système en encorbellement
buckling	Knicken	flambeage
bulging	Beulen	voilement
to restrain	einspannen	encastrer
compression flange	Druckflansch	membrure comprimée
web buckling	Ausknicken des Steges	flambeage de l'âme
direct bearing stress	reine Auflagerspannung	contrainte pure aux appuis
rust prevention	Vorbeugemaßnahmen gegen Rost	mesures antirouille
to be all for something, (to support something)	dafür sein	être pour quelque chose
to put up an argument against something, (to oppose something)	gegen etwas Einwände machen	formuler des objections contre quichi
shipbuilding dock	Baudock, Trockendock	cale sèche
mechanical lift for ships	Schiffshebewerk	ascenseur à bateaux
launching	Stapellauf	lancement
slipway	Schiffsaufschleppe	cale de halage
no end, enormously	bedeutend	sans fin, énormément
reinforcing bar	Bewehrungsstab	barre d'armature
to get a hold (to take control)	kontrollieren	contrôler
phosphating	Bondern	bonderisation
stove-enamelling	Emaillieren	émaillage
exterior quality paint	hochwertiger Außenanstrich	peinture extérieure de haute qualité
to be up against it, (to be in trouble)	in Schwierigkeit sein	être en difficulté
galvanised steel	verzinkter Stahl	acier galvanisé
enamelled steel	emaillierter Stahl	acier émaillé
plastics	Kunststoff	matière plastique
external cladding panel	Außen(wand)verkleidung oder -Behang	revêtement
protective coating	Schutzschicht	couche protectrice
patina	Patina	patine
fire protection	Brandschutz	protection contre le feu
fire hazard	Brandgefahr	risque d'incendie
fire-resistant	feuerhemmend bis feuerbeständig	résistant au feu
refractory	feuerfest, hochfeuerbeständig	réfractaire

non-inflammable	nicht entzündlich	ininflammable
gypsum plasterboard	Gipskartonplatte	panneau en carton-plâtre
asbestos cement sheeting	Asbestzementplatte	panneau en amiante-ciment
quilting	Dämm-Matteneinbau	mise en place des matelas isolant.
protective asbestos	Asbestschutz	protection en amiente
(cohesive) binder	Binder, Bindemittel	liant
gun	Spritze	canon
gunnite technique	Spritzbetonverfahren	méthode de béton projeté
plated	plattiert	dallé
plastic coated	kunstoffüberzogen oder -beschichtet	stratifié en plastique
lightweight cold formed mild steel section	Stahlleichtprofil, Kaltprofil	profile léger en acier doux
spot welding	Punktschweißen	soudage par points
self-tapping screw	selbstschneidende Schraube	vis taraudeuse
lattice truss	Fachwerkbinder	ferme en treillis
dead load	Eigengewicht	poids propre
live load	Verkehrslast	charge courante
ratio	Verhältnis	rapport
axial load	axiale Belastung	charge axial
box section	Kastenquerschnitt	section transversale du caisson
torsion resistance	Torsionswiderstand	résistance à la torsion
metal recovery	Metallzurückgewinnung	recylage du métal
to melt down	(herunter)schmelzen	réduction par en fonte
weld metal	Schmelzmetall	métal fondu
the mind boggels, (one is astonished)	man wundert sich	on s'étonne
in leaps and bounds, (rapidly)	sprunghaft, in großen Sprüngen	par sauts
to wind up, (to end, to close)	abschließen, beenden	régler
to glue	kleben	coller

5. STRESSES AND STRAINS

The structural analysis of a given structure usually begins by determining the forces acting on the static system. As with planning, structural analysis helps to find the most suitable, as well as economical form of construction which will provide adequate support and also resist all forces.

Forces

The forces likely to effect a structure are distinguished by the way in which they act, i.e. vertically, horizontally or obliquely. These forces can be either superimposed loads, such as the active earth pressure behind a retaining wall, or dead loads, an example of which is the weight of the retaining wall itself. Structural dead loads are simply calculated from the actual or estimated sizes of the members and their unit weights. A typical example of a load table for a reinforced concrete floor is given as follows: –

Dead load		400 kp/m^2	4 kN
Floor finish		100 kp/m^2	1 kN
Live load		200 kp/m^2	2 kN
	Total load	700 kp/m^2	7 kN

The task of a civil engineer is to prevent these forces from moving the structure. This is achieved, when all the active external forces and moments are counteracted by the inner forces and moments of the structure itself. Once these conditions are satisfied, the structure remains in equilibrium, or at rest in other words.

Fig. 13 shows the action of the forces exerted on an abutment and adjoining truss bridge. Under traffic, the members of a truss bridge are affected by different forces. Some forces tend to pull, whilst others exert a thrust at the junction points. The members which are pulled are termed ties and those which are pushed are known as struts. In order to design the members, the engineer must ascertain the position at which a standard train of loads rolling across the bridge has the greatest effect on the members.

Besides the normal forces which pull and exert thrust, there are shearing forces which act vertically to the action line of the first mentioned forces. If the weight of the members themselves is disregarded, and furthermore, it is assumed that the system points have no restraint, then the members will only be affected by normal forces. Where the lower chord is affected by the traffic load itself, this will naturally produce additional shearing forces and bending moments. Horizontal forces often result from traffic braking, whilst other forces are caused by the wind, which is, incidentally, an essential factor in the dimensioning of tall slim buildings like towers etc., or by the impact of ships on the embankments.

Braking forces are derived from the vertical traffic loads and are expressed as a percentage thereof, impact forces are taken into account by a special factor and wind forces are established by the Standards in relation to the structure height, the inclination of the affected area and the type of structure. Snow load is an important part of vertical force, which is also established by the Standards of the particular region. With the exceptions of the mountainous regions, the Standard is 0.75 kN/m^2 in Germany, for example, whereas in Britain, as would be expected, it is only about 0.68 kN/m^2.

Shearing Force (S.F.) and Bending Moment (B.M.) Diagrams

Fig. 14 shows the shearing force and bending moment diagrams of a simple supported beam. This beam has a uniformly distributed load, often referred to as U. D. L., between two supports, and a concentrated load on the cantilevered end. To obtain the values of the ordinates of the curves, it is usual, first of all, to calculate the support reactions. After this, the values of the ordinates of the shearing force at each load changing point can be obtained, by simply adding together all the vertical forces up to the particular section of the beam. By international agreement, the shearing force value is a positive one if the left hand side of

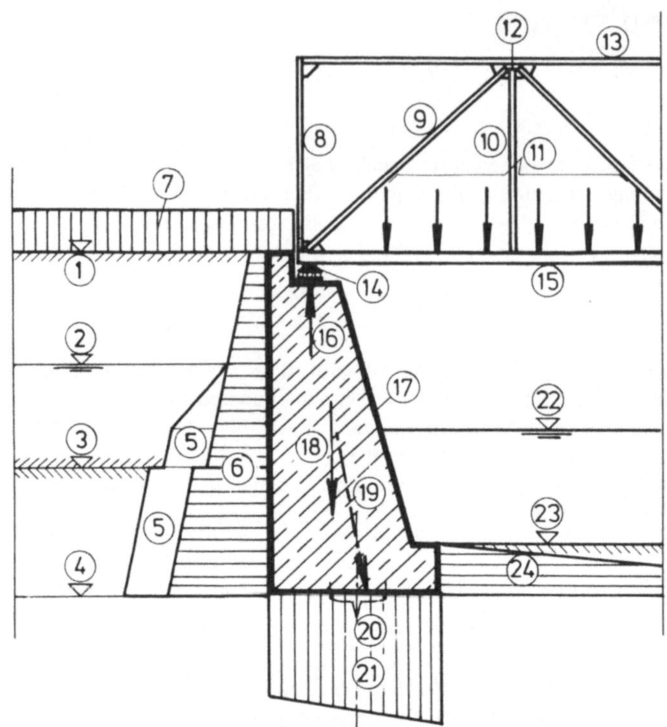

Fig. 13: Forces exerted on an Abutment and a Railway Bridge

1	**top level of ground**	Ordinate Oberkante Gelände (O.K.G.)	ligne du terrain
2	**ground water table, ground water level**	Ord. Grundwasserspiegel	nappe phréatique
3	**level of soil change**	Ord. Bodenwechsel	niveau de transition
4	**bottom of bridge abutment**	Unterkante (U.K.) Brückenwiderlager	bord inférieur de la culée
5	**(area of) resultant water pressure**	Wasserüberdruck(fläche)	surface de surpression
6	**active earth pressure**	aktiver Erddruck	poussée active du terrain
7	**traffic area load**	Verkehrsflächenlast	charge de surface de la circulation
8	**zero (force) member of the truss bridge**	Nullstab der Fachwerkbrücke	barre zéro du treillis
9	**diagonal member, strut**	Diagonal(druck)stab	barre diagonale
10	**tension member, tie**	Zugstab	barre tendue
11	**total axle loading of tender locomotive**	Lastengruppe Tenderlok	charge des essieux de locomotive avec tender
12	**system point (or centre) with junction plate**	Knotenpunkt mit Knotenblech	noeud avec gousset en tôle
13	**top chord**	Obergurt	membrure supérieure
14	**roller bearing**	Rollenlager	appui à rouleaux
15	**lower chord with track**	Untergurt mit Fahrbahn	membrure inférieure avec voie
16	**bearing pressure, bearing reaction**	Auflagerkraft, Auflagerreaktion	réaction aux appuis
17	**reinforced concrete abutment and retaining wall**	Stahlbeton-Brückenwiderlager und Stützmauer	culée en béton armé et mur de soutènement

18	**abutment dead weight**	Eigengewicht des Widerlagers	poids propre de la culée
19	**resultant**	Resultierende	résultante
20	**extent of core**	Kernweite	largeur de noyau
21	**soil resistance area**	Bodenwiderstandsfläche	surface de résistance du sol
22	**canal water level**	Ord. Kanalwasser	niveau d'eau du canal
23	**canal bed level**	Ord. Kanalsohle	niveau du fond du canal
24	**passive earth pressure**	passiver Erddruck	poussée passive du terrain

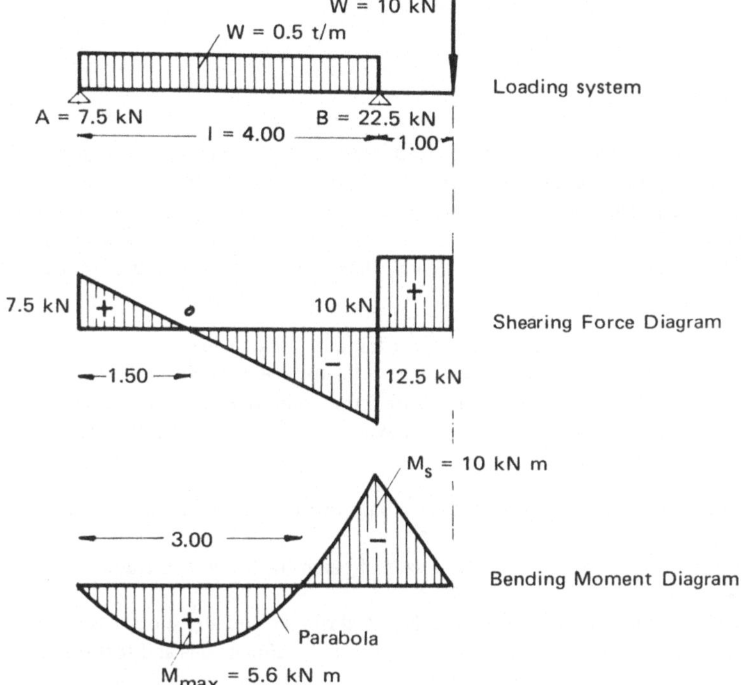

Fig. 14: Shearing Force and Bending Moment Diagrams of a Simple Supported Beam with a Cantilever

the beam, supposedly cut off, tends to move upward when activated by all the attacking forces of that part of the beam. Where this occurs on the right hand side, the position is vice versa, and so the value is a negative one.

The bending moment values of the critical points of a bending moment diagram can be found by two methods. One is to determine the sum of all the bending moments, created by the forces on one side of the section. This is similar to the method of determining the support reaction forces. Both the loads and reactions must be taken to one side of the section and the moments taken about this section. The other method is to add together all the areas formed by the shearing force curve and the zero line up to the point in question, taking into account the positive or negative signs.

Referring to Fig. 14, the following facts can be summarised. They are: –

(a) that a U. D. L. produces a uniformly sloping S.F. diagram, whereas concentrated loads produce a vertical drop in the S. F. line for every point,
(b) that a U. D. L. produces a parabolically shaped B.M. diagram, whereas each concentrated point loading forms a sharp peak in the B. M. line, and
(c) that the extreme B.Ms., i. e. max M and min M, correspond to the zero points of the S. F. curve on the base line.

Bending Curves and Influence Lines

There are other significant curves in the field of structural analysis, namely bending curves and influence lines. Bending angles and curves can be obtained in principle by taking the area formed by the B. M. curve as load. The B. M. on the system resulting from this load is then equivalent to sagging.

Influence lines show the value of a structural analysis unit, such as a support reaction, the S. F. or the B. M. on a certain point of the structure, for a load wandering across the system, e. g. a beam. The value is given directly under the relevant position of the load. Influence lines are important in calculating for bridges, where the influence of rolling traffic loads is very significant.

Complex Loading Conditions

Complex loading conditions have often to be dealt with, and in such cases, it is preferable to apply the method of superposition, which entails breaking down the load system into a series of manageable loading systems. Firstly, the required force actions for each simplified system must be determined. Then it is necessary to schedulise and add together all the values resulting from load situations which might exist at the same time. When considering all possible loads, those caused by the actual process of construction should not be overlooked.

Rather than become involved in long, tedious and often complicated manual calculations, engineers often employ electronic digital computers nowadays. In as little time as a minute, or even less, a computer can produce the single values resulting from one load, or more commonly from a simple load composition, of practically all the normal construction systems. An experienced operator is able to feed the computer with four kinds of information. These are: –

(a) a punched programme card or strip, or alternatively a magnetic tape or similar form of information media,
(b) the dimension, e. g. spans and section sizes of beams or frame construction,
(c) the loads and
(d) the material constants, i. e. the modulus of elasticity and the moments of inertia, if the computer is unable to calculate the latter from the given dimensions and section sizes.

Economics invariably dictates whether or not it is feasible to employ a computer. There is always a danger that the machine may be too large, and consequently expensive, when it is designed to cope with all the necessary calculations involved in structural analysis.

Hyperstatic Systems

For comparatively simply and commonly applied statically indeterminate (or hyperstatic) systems, such as beams or slabs running continuously over a number of unyielding supports, there are some well proven quick calculation methods, which require no machine, but are well within the scope of a slide rule.

In the case of a continuous beam resting on more than two supports Clapeyron's Theorem of Three Moments assists the engineer in ascertaining values with different spans and loads, but preferably with a constant moment of inertia along the beam. In instances where the spans are equal, or nearly so, there are tables available for multi-span beams and for all normal loads.

When dealing with frames, the B. Ms. at the joints and supports, usually being two, can be successfully determined with methods of moment distribution. The method of Hardy Cross or Kani, which embodies this principle, is commonly adopted for this analysis. By iteration, according to Cross, the initially assumed static values will be corrected step by step, by moving through the system and gradually reaching the exact B. M. values with each stage, thus arriving at the correct values.

At first, all joints are assumed to be completely restrained. The B. Ms. resulting from the loaded members restrained at the joints will be corrected in such a way, that only one joint will be released at a time, whilst the others are restrained. This process is repeated until all the B. Ms. existing on each joint will be adjusted, The inserted and remaining moments at any one calculation process are distributed, according to the stiffness of the members which form the joint. By a so called carry-over factor, the distribution moments are transferred to the opposite side of the members.

For design using the iteration analysis, the following process must be carried out: –

(a) the determination of the restrained B. Ms. in each member using the appropriate tables,
(b) the calculation of the stiffness of the members,
(c) the calculation of the distribution factors for each joint and
(d) the determination of the moment distribution, taking into account the carry-over factors.

Internal Forces and Moments

Whatever static methods are employed, the results are always the values of the following: –

(a) the normal forces pulling or pushing in the longitudinal axis of a member,
(b) the shearing forces acting at right angles to the longitudinal axis,
(c) the bending moments and
(d) the torsion, which may cause a twisting action about the longitudinal axis of a member.

Members which are uniform in section and in their material content throughout their entire length, e. g. steel or timber beams, must be designed for the maximum static values. The size of the section and reinforcement of reinforced concrete members, such as beams or columns, can be suited to the varying stresses and strains. In most cases, not only one but a combination of some of the stresses and strains just listed in categories (a) to (d), e. g. pressure and bending, may act on a member at the same time. All combinations are possible and commonly occur, even double bending, which is bending in the direction of two axis.

Design – Permissible Stresses

Having completed the structural analysis, good design must logically follow, whereby members of sufficient strength, within reasonable limits, are found to resist the attacking forces and stresses and strains. The permissible stress in a member is normally the failure stress divided by the safety factor. This factor varies from 1.1 to 5.0, according to the degree of homogeneity of the material, and the quality (e. g. of timber), plus the loading category.

To determine the internal forces which must resist the external ones, the strength of the material and the moment of resistance of the critical sections of the member in question must be considered. Different materials have different permissible safety limits for various stresses and strains. In the case of inhomogeneous or composite materials, like timber or reinforced concrete respectively, the values for the permissible stresses are even dependent on the directions in which the attacking forces act. For example, the permissible stress for timber, in which the grain runs parallel to the longitudinal axis – in other words its permissible shearing stress acting parallel to the wood fibres – is only about one tenth of its permissible stress for compression or tension, i. e. normal forces in the longitudinal axis of the member. Although there is some resistance strength against pulling in plain concrete, it is so minimal that it is usually ignored. All pulling forces in concrete must be resisted by the reinforcing bars, whilst pressure resistance is in the concrete itself.

Moments of Resistance

Dealing with moments of resistance generally presents little difficulty, since there are standard tables, even in elementary text books on stresses and strains, which provide the values for practically all sections of structural members. The values for all the normal timber members, such as purlins, rafters, joists, struts etc., and the more common steel members, like stanchions, columns, box beams, channels, angles and tees, are readily available. For the more complex sections, which are either made up of two or more geometrical areas, or are the usual industrially produced profiles, like H- or Z-beams, it is relatively simple to calculate the moment of resistance for the two main axis. These are the ones which give the highest and lowest values.

For centrally loaded stanchions it is preferable to have a section possessing two approximately equal moments of resistance. If standard tables do not give such a section with the required loading capacity, because a standard section is not strong enough, a combination of several standard sections of sufficient strength is advisable. Fig. 15, for example, shows a section through such a composite steel stanchion with a substance-free centroid. Another example of a built-up member is often provided by a large beam. Fig. 16 shows a section through a lightweight compound beam of laminated timber glued together. In the case of beams, it is more effective to have a section with one prominent moment of resistance. In both cases, the moments of resistance have to be related to the main axis of gravity of the combined section, and this should be done in schedule form. To calculate this for a combined section, it is necessary to know the area, the centre of gravity and the moment of inertia for each individual section.

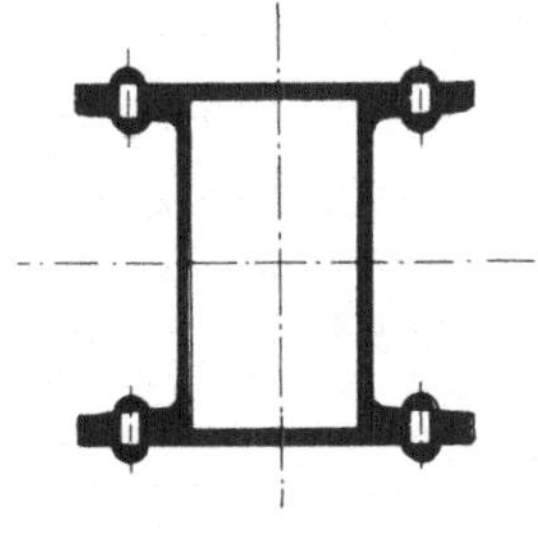

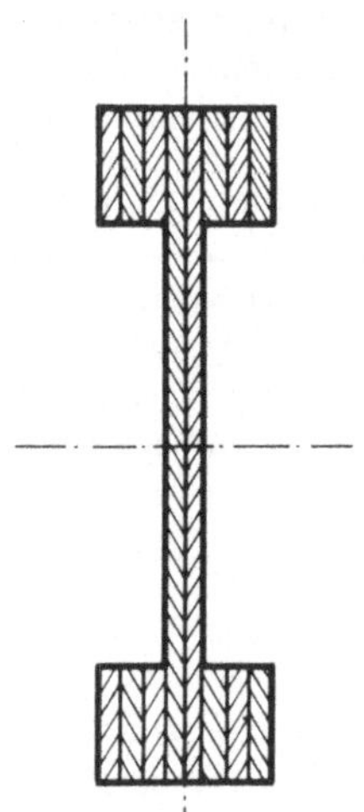

Fig. 15: Section through a Composite Steel Stanchion

Fig. 16: Section through a Laminated Timber Beam

Moments of Inertia

All the moments of inertia, as well as the moments of resistance, are in no way connected with the material of the member. These are just mathematical values, which are only dependent on the sectional area of the member. In modern timber construction, the frequently employed compound section should be regarded as a unit. This means that their components must be securely connected, so that the joints are at least as strong as the components themselves.

An example of the properties and stresses of a rectangular section of a homogeneous material is given in the following table: –

Section	Area	Moment of Resistance or Section Modulus	Moment of Inertia	Working Stress in a pulled or pressed short member	Bending Stress in a Beam
y, b, d, X–X, y; Neutral Axis; Centre of Gravity	A = b d	$Z_{xx} = \frac{b\,d^2}{6}$ $Z_{yy} = \frac{d\,b^2}{6}$	$J_{xx} = \frac{b\,d^3}{12}$ $J_{yy} = \frac{d\,b^3}{12}$	$f = \frac{W}{A}$ (W is the total load on the section)	$f_b = \frac{Z}{M}$ (M is the bending moment)

The formulae given in this table are based on Hooke's law, which states that within the elastic limit of a member, the stresses and strains are proportional to the imposed loads. As applied to the longitudinal fibres in a bent beam, it means that stress in a beam fibre is linear to its strain. Not only the strain variation diagram, but also the stress variation diagram, of a rectangular section of a bent beam will therefore be linear in character.

Deflection

All suspended forms of construction are designed in accordance with their characteristic values of the B. M. s and S. F. s. There is, however, another value that has to be considered, and this is maximum deflection. Several well proven Standards exist, which prescribe the utmost permissible deflection for different types of construction and for varying sizes, spans and materials. Normally, the highest permissible deflection for steel or timber beams is between 1/200th and 1/500th of the span. For similarly supported reinforced concrete slabs and beams, however, the degree of slenderness – that is the ratio $\frac{l}{h}$ (span/depth) – should not exceed 35 up to a span of 4.50 metres. A section of any loaded member is most suitable and economical when it comes just within the limits of stress and deflection of any member under different support and load conditions. For a simple supported beam carrying a U. D. L. of W kp/m, with the span l and the well known B. M. max equal to $\frac{w\,l^2}{8}$, the deflection is as follows: –

$\frac{5}{348}\,\frac{w\,l^4}{EJ}$, where E is the modulus of elasticity (Young's modulus) of the material and J is the moment of inertia of the section.

Buckling

In the calculation and design of a column, the tendancy for the member to buckle sideways must be fully taken into account. Besides the properties of the section and the material, the slenderness ratio and the nature of the end supports are important. The old established Euler's formula for crippling stress is still in use today, and it is as follows: –

$$\frac{W}{A} = \frac{E\pi^2}{\left(\frac{l}{r}\right)^2}$$

Where W is the single axial load on top of the column,
A is the cross sectional area,
$\frac{l}{r}$ is the slenderness ratio,
l is the effective length
and r is the radius of gyration.

The influence of the nature of end supports, whether a column or strut is held in position at one or both ends, must be taken into account by the factor with which the actual length has to be multiplied to get the effective length. Apart from this general analysis method, other forms of calculation have been developed, primarily in respect of the characteristics of the material from which columns and struts are made.

As an example, the axially loaded reinforced concrete column should be considered here. There are two outstanding forms of column, one being rectangular in section and having reinforcing stirrups, and the other being circular in section and having helical binding reinforcement.

In short columns, the B. S. (British Standard) for example, states that the permissible load is as follows: –

$P_O = p_{cc}\,A_c + p_{sc}\,A_{sc}$
where p_{cc} is the permissible stress for concrete in direct compression,
A_c is the cross sectional area of concrete,
p_{sc} is the permissible compression stress for steel reinforcing rods in columns and
A_{sc} is the cross sectional area of the longitudinal steel reinforcement.

A short column is regarded as one where the ratio of $\frac{\text{effective length}}{\text{least lateral dimension}}$ equals or is less than 15. In respect of short columns, the spacing of the reinforcing stirrups should not exceed at least (a) the smallest lateral dimension of the column section, (b) 12 times the diameter of the smallest longitudinal reinforcing bar, and (c) a distance of 30 cms.

When considering long columns, the permissible stress or the applied load must be reduced by a factor, which depends on the grade of slenderness, as given in the relevant table. For columns reinforced with helical binding, there is an additional formula, which also takes into consideration the steel of the helix.

Eccentric Loading

Eccentric loading has often to be dealt with in vertical members, like pillars, piers, columns, or the legs of a framed structure, where any of these incorporate a short cantilevered member for carrying a travelling crane rail etc. in a factory or workshop. The analysis and design of this type of member is that adopted for one which is pressed and bent at the same time. If constructed in reinforced concrete, the quantity of tension bars necessary for the B. M. can often be reduced through the influence of the additonal pressure caused by the thrust.

Foundation Design

Foundation design is comparatively simple, if only vertical forces are present. It means that the base area of the construction, whether it is a building or a bridge etc., must be large enough to spread the load and thus avoid excess settlement. In other words, the permissible soil pressure must not exceed a critical value. The determination of the vertical loads, resulting from the construction and from possible traffic or other load affecting influence, is usually straight forward, but nevertheless, experience and care is necessary to ascertain the soil resistance. This task is of a specialised nature, which forms part of soil mechanics, entailing site investigation with drilling rigs, etc., followed up by sample testing in a research laboratory.

In dimensioning foundations, it is necessary to consider more than just the direct pressure on the soil. The pressure on the deeper soil strata should also be taken into account within a certain distribution angle, especially where they are likely to be of poor bearing capacity. With cohesive soils, extreme care in foundation design is essential, mainly in respect of their water content. Their resistance against superimposed loads depends largely on the extent of this water content, which is responsible for the plasticity grading of this type of soil. Water must, therefore, be prevented from infiltrating into such soil under foundations. Although the probable settlement of a foundation can only be very roughly calculated, because of so many uncertain factors, nevertheless the settlement value should always be a decisive factor in the calculation and design of a foundation.

Where foundations are subjected to both vertical and horizontal forces, it is necessary to investigate the position of the stresses at the junction of the foundation construction and the soil at its base. The Middle Third Rule states that there is no tension stress in the base section when the resultant force falls within the middle third or core. Where the resultant force passes exactly through a point on the core perimeter of the middle third, the edge of the base will have a stress value of zero. Even if the resultant is within the core, the necessary stability of the construction cannot be guaranteed. As a result, the following investigation is essential: –

(a) that the permissible soil pressure is not exceeded at any edge, and that safeguards are taken in this respect against soil shear failure,
(b) that the friction between the foundation base and the soil is sufficient to counteract any sliding effect caused by horizontal forces,
(c) that the safety margin against the construction overturning, caused by a possible tilting moment, is high enough and
(d) that there is an adequate safety margin against soil shear failure, which results in the tendancy for a large segmental volume of soil to move.

The continual persuance of structural analysis in relation to stresses and strains, amongst other aspects, is essential, in order to make better use of existing materials and techniques, and to ascertain how to utilise new materials and ideas to the greatest advantage. Many of the long established laws and formulae will doubtless remain for years to come, but fresh ones will inevitably emerge to meet the demands of newly developed materials and shapes of up-to-date construction.

Vocabulary	Wörterverzeichnis	Glossaire
stress	Spannung	contrainte
strain	Dehnung	élongation
stresses and strains	Beanspruchungen, Statik	sollicitation, statique
structural analysis	Baustatik	statique de constructions
structure	Bau, Bauwerk	ouvrage
support	Unterstützung, Auflager	appui
force	Kraft	force
to superimpose	überlagern	superposer
superimposed load	Auflast, Nutzlast	surcharge, charge utile
active earth pressure	aktiver Erddruck	poussée active du terrain
retaining wall	Stützmauer	mur de soutènement
dead load	Eigengewicht	poids propre
unit weight	Raumgewicht	densité
load table	Zusammenstellung der Lasten	tableau de charges
reinforced concrete floor	Stahlbetondecke	plancher en béton armé
live load	Verkehrslast	charge courante
floor finish	Fußbodenbelag	revêtement de sol
abutment	Widerlager	culée
truss bridge	Fachwerk(träger)brücke	pont en poutres en treillis
member	(Teil-)Glied, Stab	barre
thrust	Schub	poussée
tie	Zugstab	tirant
strut	Steife, Stütze, Druckstab	étrésillon
standard train of loads	genormter Lastenzug	train de charge normalisé
shearing force	Scherkraft	cisaillement
system point	Knotenpunkt	noeud
to restrain	einspannen	encastrer
chord	Gurt	membrure
bending moment	Biegemoment	moment fléchissant, moment de flexion
braking	Bremsen	freinage
dimensioning, sizing	Bemessen, Bemessung	dimensionnement
impact	Stoß	choc
embankment	Kaje	remblai
Standards	Normen	normes
simple supported beam	Balken auf 2 Stützen	poutre sur deux appuis
uniformly distributed load (U.D.L.)	Gleichlast	charge uniforme
concentrated load	Einzellast	charge isolée
cantilevered end	Kragarm	porte-à-faux
support reaction force	Auflagerreaktion, Auflagekraft	réaction aux appuis
zero line, base line	0-Linie, Grundlinie	base
sign	Vorzeichen	signe
multi-span beam	Träger auf mehreren Stützen	poutre sur plusieurs appuis
moment distribution	Momentenverteilung	distribution des moments
bending diagram curve	Biegelinie	ligne fléchie
bending angle	Drehwinkel	angle de rotation
sagging, deflection, deflexion	Durchbiegung	fléchissement, déflexion

structural analysis unit	statische Größe	valeur statique
superposition	Überlagerung	superposition
to break down – to divide into smaller parts	in Abschnitte teilen	diviser
load system, loading system	Belastungsfall	cas de charge
manual	von Hand, ohne Maschinenhilfe	manuel
electronic digital computer	Digitalrechner, Computer	calculateur digital
punched programme card	Lochkarte	carte perforée
magnetic tape	Magnetband	ruban magnétique
span	Spannweite	portée
frame	Rahmen	cadre
modulus of elasticity	Elastizitätsmodul	module d'élasticité
moment of inertia	Trägheitsmoment	moment d'inertie
statically indeterminate, hyperstatic	statisch unbestimmt	isostatique
slab	Platte	dalle
unyielding support	festes Auflager	appui fixe
slide rule	Rechenschieber	règle à calcul
continuous beam	Durchlaufträger	poutre continue
multi-span beam	Träger auf mehreren Stützen	poutre à appuis multiples
moment distribution	Momentenverteilung	repartition des moments
iteration	Wiederholung	répétition
to release	loslassen	lacher, libérer
to adjust	ausgleichen	ajuster
stiffness of the member	Stabsteifigkeit	rigidité de la barre
carry-over factor	Übertragungsziffer	facteur de report
restrained B.M.	Einspannmoment	moment d'encastrement
distribution factor	Verteilungszahl	coefficient de distribution
longitudinal axis	Längsachse	axe longitudinal
twisting	Verdrehen, Verdrehung	torsion
steel beam	Stahlträger	poutre en acier
timber	Bauholz	bois de construction
reinforcement	Bewehrung	armature
column	Stütze	poteau
permissible stress	zulässige Spannung	contrainte admissible
failure stress	Bruchspannung	contrainte de rupture
safety factor	Sicherheitsfaktor	coefficient de sécurité
loading category	Belastungsfall	catégorie de charge
moment of resistance	Widerstandsmoment	moment de résistance
composite material	Verbundstoff	matériau composite
grain, fibre	Faser	fibre
permissible shearing stress	zul. Scherspannung	contrainte de cisaillement admissible
compression	Druck, Zusammendrückung	compression
tension	Zug	tension
resistance strength	Festigkeit	résistance
plain concrete	unbewehrter Beton	béton non armé
pulling force	Zugkraft	force de traction
reinforcing bars reinforcing rods	Bewehrungsstäbe	barres d'armature
pressure resistance	Druckfestigkeit	résistance à la compression
standard table	Normtafel, Tabelle mit Richtwerten	tableau de valeurs standard
purlin	Pfette	panne
rafter	Sparren	chevron
joist	Balken	poutre
strut	Strebe	étrésillon

stanchion	Stahlstütze	poteau en acier
box beam, box girder	Kastenprofilträger	poutre-caisson
channel	U-Profil	profilé en U
angle	Winkel-Profil	cornière
tee	T-Profil	profilé en T
H-beam	Breitflanschträger	poutre à ailes larges
Z-beam	Z-Profilträger	poutre en Z
standard section	genormtes Profil	profilé normalisé
substance free centroid	stoffreie Achsen	axe supposé
built-up member	mehrteiliges (Bau-)Teil	élément composé
compound beam	Verbundträger	poutre composite
laminated timber	Schichtholz	bois lamellé
section modulus, modulus of section, moment of resistance, (Z)	Widerstandsmoment, (W)	moment de résistance
working stress	Gebrauchsspannung	tension de travail
bending stress	Biegespannung	contrainte de flexion
elastic limit	Elastizitätsgrenze	limite d'élasticité
imposed load, superimposed load, applied load	Auflast, Nutzlast	surcharge, charge utile
longitudinal fibre	Längsfaser	fibre longitudinale
bent beam	Biegeträger	poutre fléchie
slenderness	Schlankheit	élancement
to buckle sideways	ausknicken	flamber
slenderness ratio, $\frac{l}{r}$	Schlankheitsgrad $\frac{s_k}{i}$	degré d'élancement
crippling stress	Knickspannung	contrainte de flambeage
single axial load	mittige Einzellast	charge isolée axiale
effective length, (l)	Knicklänge, (s_k)	longueur de flambeage
radius of gyration, radius of inertia, (r)	Trägheitsradius, (i)	rayon de giration
actual length	geometrische Länge	longueur réelle
axially loaded	mittig belastet	chargé au centre
reinforced concrete column	Stahlbetonstütze	poteau en béton armé
stirrup	Bügel	étrier
helical	schraubenförmig	hélicoïdal
helical binding reinforcement	Umschnürung	frettage
permissible load	zul. Belastung	charge admissible
direct compression	reiner Druck	pression pure
permissible compression stress	zulässige Druckspannung	contrainte de pression admissible
longitudinal steel reinforcement	Längsbewehrung	armature longitudinale
eccentric loading	außermittige Belastung	charge excentrique
framed structure	Rahmenbauwerk	ouvrage en cadre
cantilevered	ausgekragt	en porte-à-faux
travelling crane	Laufkran	grue roulante
rail	Schiene	rail
factory	Fabrik	usine
workshop	Werkstatt	atelier
tension bar	Zugstab	barre tendue
foundation	Fundament, Gründung	fondation
base area	Grundfläche	surface de base
to spread the load	Last verteilen	distribuer la charge
settlement	Setzung	tassement
permissible soil pressure	zul. Bodendruck	pression admissible sur le sol
soil resistance	Bodengegendruck, Bodenpressung	résistance du sol

soil mechanics	Bodenmechanik	mécanique du sol
site investigation	Bodenuntersuchung	investigation du sol
drilling rig	Bohrausrüstung, Bohrgerät	équipement de forage
sample testing	Untersuchung von Proben	examen des éprouvettes
research laboratory	Forschungs-Laboratorium	laboratoire de recherches
soil strata	Bodenschichten	couches de sol
distribution angle	Verteilungswinkel	angle de distribution
bearing capacity	Tragfähigkeit	capacité de charge
cohesive soil	bindiger Boden	sol cohésif
water content	Wassergehalt	teneur en eau
plasticity grading	Bildsamkeitsgrad, Zustandsform	indice de plasticité

Dialogue

Robert: Do you feel that too much emphasis and effort are put into theory in today's construction work?

Duncan: Well, theory certainly played a far smaller part in years gone by, when many magnificant buildings, like cathedrals, were constructed. In fact, practical experience was the main basis of the architect's design. Mind you, there must have been a great deal of trial and error in those days, particularly with the notable civil engineering works.

Robert: Even today, you know, some architects avoid theory involving precise calculations and design, and as a result, we often get overdimensioned members and the wasteful use of materials, in my opinion.

Duncan: On the other hand, of course, there are those "handbook designers" who sometimes misuse theory.

Robert: I take it you mean that they carefully select the individual members of adequate section for the job, but combine them badly in the actual structure.

Duncan: Yes, indeed, and serious failures can so often occur as a result.

Robert: I believe that the perfect designer, if such a man exists, is one who has gained much practical experience and also has a wide knowledge of theory. This may be asking rather a lot, but a balanced combination of some practice and theory is necessary in all designers.

Duncan: It should be remembered, of course, that a sound theoretical understanding of structures is required to make full use of the many Standards and Codes of Practice.

Robert: These so-called Standards etc. do, of course, give much useful information in the form of basic tables and details, but as you have already implied, they need to be in the right hands. It's interesting to note that there is a growing tendency in some forms of construction to put more reliance in standard tables and details.

Duncan: Could you say a little more about this, Robert, if possible?

Robert: By all means. Well, with skeleton framing, for example, for such buildings as skyscrapers, meticulous theoretical calculation is often abandoned in favour of the established tables etc.

Duncan: For what reason, precisely.

Robert: It's simply this. Since the failure of one single foundation member, through undue settlement, can alter the bending moments in the connected members, even from a plus to a minus value, detailed calculation is unrealistic, and therefore hardly seems justified.

Duncan: I can understand this, for the behaviour of the soil is so uncertain. Incidentally, I feel more consideration should be given to the peculiar characteristics of some soil strata.

Robert: Where standard tables can't provide the right answers, other methods, like that of Hardy Cross, Professor of the University of Illinois, have to be adopted. This ingenious method, as you probably know, is used for analysing statically indeterminate systems.

Duncan: You did mention to me once in passing, that you were part of a consulting engineer's team which actually adopted this method of successive approximations, when it was first introduced into Europe from America.

Robert: That's quite true, and believe me it was an important event. Every civil engineer, I feel, should really understand this method, even if he has no great knowledge of higher mathematics. However, it does take time to change over to Cross, especially for my older colleagues, who are so familiar with the more conventional mathematical methods.

Duncan: Since you are experienced in this method, perhaps you would give your opinion as to what is so different with the conception of Cross.

Robert: Well, let's take, for example, a continuous frame with fixed supports. This is, as you know, highly statically indeterminate, and it takes a long time using old methods to arrive at the static values which are necessary for structural design.

Duncan: Sorry to butt in, but in cases like this, surely the long established angle balancing method could be useful.

Robert: Yes, I agree, but with Cross, who is aiming to get an equilibrium of the moments in the joints and not of the angles, the process in most cases is so much easier and quicker.

Duncan: In the beginning of the process, one has to determine all the moments at the ends, as if they are fully restrained. How do you manage this?

Robert: Actually, these can be found from the many tables for all possible loads. Just imagine how quickly it can be done for each end of a member. The sum of these moments, taking into account whether they are plus or minus, should in the end give you zero. At this stage the unbalanced moment is large, and the next step is to distribute this rest moment amongst the connecting moments in proportion to their stiffness.

Duncan: Is this so-called stiffness the moment that has to be applied at one end of a member to produce a unit rotation of that end, when the other end is fixed?

Robert: Yes, that is so, but we never speak of moments, but of k constants. This is correct, because the E-Module can be left out where only a single material is involved.

Duncan: Now then, having multiplied the rest moment by the different k constants, one has to carry over these moments moving away from the joint. How exactly do you do that?

Robert: This is also very easy. The carry-over factor is 0.5 in the normal case of a uniform moment of inertia. One has only to carry over half of the moments moving away from the joints to the other end of the member.

Duncan: I see. Then I believe you distribute these moments just carried over, and repeat the process until they are small enough to be neglected.

Robert: Yes, precisely. The resultant moment is arrived at by adding together the fixed-end moments, distributed moments and the carried over moments. That's all there is to it.

Duncan: Not as sophisticated a system as many people think. Nevertheless, doesn't the Cross method have its limitations?

Robert: Not really, but there are, however, other similar methods. The Kani method, which we briefly mentioned in our chapter, has advantages in certain cases.

Duncan: Could you give such a case?

Robert: Well, in all systems where horizontal members are displaceable, the Kani method is particularly suitable. Take, for example, multi-storey portal structures. Here, the calculation of the moments at the joints, including the compensation moments against displacement, are determined in one go using Kani.

Duncan: I think it should be said that both the Cross and Kani methods are not just approximate ones. If the process is repeated often enough, the results can be absolutely correct.

Robert: As a complete change, could we now focus our attention on the subject of timber construction using glued laminated components?

Duncan: Why not, for it's sure to be of some interest to many of our readers. It's right up your street, so perhaps you'd make a start.

Robert: Fair enough, Duncan. I'll begin by posing you a question.

Duncan: Please go ahead, Robert.

Robert: Well, what is your reaction to recent developments in this form of timber construction?

Duncan: You already know, I'm sure, that we don't make such full use of laminated timber in Britain as you do on the Continent, but nevertheless, I believe it has distinct possibilities in the appropriate conditions.

Robert: I'm rather surprised that you in Britain haven't jumped at the chance to make greater use of such a versatile material.

Duncan: Not really so surprising, since there's only a limited amount of good seasoned timber available in Britain for this purpose. Most of our timber has to be imported, and at the moment the cost is rather high.

Robert: Anyway, to get back to the point, I should say that laminated timber construction requires experience and care in manufacture. The timber must be specially selected, having particular regard to it's moisture content. The surface too must be free from oil and dust and should be smooth.

Duncan: We ought to mention the glue as well, didn't we?

Robert: Yes, indeed. It too must possess special qualities to provide adequate strength under stress and for the anticipated life of the member.

Duncan: I'm sure that the significant advance in the quality of modern glue has given timber a new lease of life.

Robert: To a large extent, this is perfectly true. Even for the wide spans employed in sports pavilions, exhibition and congress halls, factories and workshops and many other kinds of communal assembly building, glued laminated timber members are becoming increasingly popular here on the Continent.

Duncan: Years ago, of course, nobody would even have dreamt of using such lightweight members in these situations. Steel or reinforced concrete would certainly have been the order of the day, even if unjustified.

Robert: Having strayed a little from our actual subject, I think we should now concentrate on more relevant items.

Duncan: Maybe you're right, Robert, and as Stresses and Strains is more your subject than mine, I leave it to you to lead the way.

Robert: Just as you like, Duncan, and I suggest we discuss soil pressure for a start.

Duncan: I go along with that all right, because I rate this particular topic as being very important.

Robert: As you know, much has already been said by many experienced engineers about this, because there are numerous problems with soil pressure in relation to retaining walls, sheet piling and similar constructions.

Duncan: Could you give a little more detail, please?

Robert: Yes, I'll be glad to. Take, for example, the problem existing behind a normal rigid retaining wall. Here, the diagramatic shape of the active earth pressure acting against the wall is triangular. The force of earth pressure is inclined to the rear of the wall, and the angle largely depends on the roughness of the rear surface of the wall. As you know, one can replace the inclined resultant force with a vertical and horizontal force. The horizontal force tends to topple the wall over, whilst the vertical force helps to counteract this tendency.

Duncan: It follows then, that it's advisable to make the rear surface very rough to increase the vertical force and so reduce the danger of toppling.

Robert: You've actually hit the nail right on the head. By the way, have you ever come across a reinforced concrete retaining wall having a projection at the rear, rather like a rucksack on a man's back?

Duncan: I know exactly what you mean, and I think it's a good idea, which you so often make use of on the Continent.

Robert: Yes, you're right, we do. This projection, or hump, creates an effective counteraction against toppling, through the action of the soil on the hump.

Duncan: Referring back to the treatment of the rear of a retaining wall, I take it that you think the idea of making it very smooth with wrought formwork, or even treating it with bitumen or tar tanking can often prove detrimental to the stability of the wall.

Robert: In many instances, I do indeed. I'll qualify my contention though, by saying that tanking is very often necessary to resist dampness, especially where the wall forms part of a building.

Duncan: For normal retaining walls used in connection with landscaping and roadworks in hilly country, it's practical to leave the rear face rough and form weep holes in the wall to allow water to pass through, and so reduce the pressure of water on the wall.

Robert: May I at this point enlarge on the term active earth pressure?

Duncan: Please do, Robert.

Robert: Well, I won't elaborate too much, but it certainly should be stated that it depends upon the height of the earth behind the wall and the properties of the soil.

Duncan:	Can you be just a little more specific about these properties?
Robert:	Very well. Two of the main ones are its unit weight and its internal friction coefficient.
Duncan:	What about its cohesion coefficient, known as c?
Robert:	Well, this can be helpful in cohesive soil, where a safe water content is assured.
Duncan:	There is some risk then involved in applying this coefficient, and in cases of doubt, it should be disregarded.
Robert:	This is quite so.
Duncan:	Sorry to keep on picking your brains, but I'm sure our readers would benefit no end from your specialised knowledge of earth pressure diagrams. Would you care to make any comments?
Robert:	As you have virtually forced my hand with your build-up of my ability to be of assistance, I shall gladly pass on some useful information. Let me begin, first of all, by saying that the classical triangular shaped earth pressure diagram cannot be applied in every case.
Duncan:	I gather you are thinking of the deformation of this diagram, are you not, Robert?
Robert:	Yes, I am, Duncan. This occurs, as you know, where the retaining wall, or alternative means of earth support, is not rigid in character. As a matter of interest, I recently cast my eye over some calculations for temporary planking and strutting to a large deep pit. In this particular case, the diagram resembled a rectangular shape.
Duncan:	This is quite interesting, but perhaps you could say how this shape was ascertained.
Robert:	It was, in fact, ascertained by the use of pressure gauges fastened at regular intervals to the rear side of the planking.
Duncan:	Well, there we have a classic example of practice going hand in hand with theory by using empirical methods backed up later by research and calculation. I can't think of a more appropriate point at which to conclude this chapter, and so give our readers a chance of a break.

Vocabulary	Wörterverzeichnis	Glossaire
trial and error, (experiment and mistakes)	Herumprobieren	essais empiriques
civil engineering work	Ingenieurbauwerk	travaux de génie civil
Code of Practice	Richtlinien des 'British Standard Institute', (BSI)	code britannique des normes (BSI)
skeleton framing	Rahmenskelett, räumliches Fachwerk	ossature
skyscraper	Wolkenkratzer	gratte-ciel
successive approximation	schrittweise Annäherung	approximation successive
continuous frame	Mehrfeldrahmen	cadre continu
angle balancing method	Drehwinkel-Verfahren	méthode de l'angle de rotation
joint	Knotenpunkt	noeud
restrained	eingespannt	encastré
unbalanced moment	nicht ausgeglichenes Moment	moment non équilibré
rest moment	Restmoment	moment résiduel
stiffness	Steifigkeit	rigidité
unit rotation	Drehung um den Betrag 1	unité de rotation
carry-over factor	Übertragungsziffer	facteur à reporter
resultant moment	endgültiges Moment	moment resultant
fixed-end moment	Einspannungsmoment	moment d'encastrement
multi-storey portal structures	mehrgeschossiges Rahmentragwerk	cadre à plusieurs étages
compensation moment	Ausgleichsmoment	moment d'équilibre
glue	Leim	colle
glued laminated component	lamellenverleimtes Bauteil	élément lamellé
to be right up one's street, (to be one's speciality)	jemandes (Fach-)Gebiet betreffen	concernant le rayon de quelqu'un

to jump at the chance, to take advantage	die Gelegenheit ergreifen, zupacken	saisir l'occasion
seasoned timber	abgelagertes Bauholz	bois séché
moisture content	Feuchtigkeitsgehalt	teneur d'humidite
to give someone a new lease of life, (to give someone another chance)	jemandem noch eine Chance geben	donner une nouvelle chance
sports pavilion	Sporthalle	salle de sport
exhibition	Ausstellung	exposition
to be the order of the day, (to be normally used)	an der Tagesordnung sein	d'usage courant
to go along with something, (to agree with something)	mit etwas einverstanden sein	être d'accord avec
retaining wall	Stützmauer	mur de soutènement
sheet piling	Spundwand	rideau de palplanches
triangular	dreieckig	triangulaire
wrought formwork	gehobelte Schalung	coffrage raboté
tanking	Sperranstrich	enduit isolant
landscaping	Landschaftsgestaltung	aménagement des paysages
roadworks	Straßenbauarbeiten	travaux routiers
hilly country	hügliges Gelände	terrain onduleux
rear face	Hinterseite	derrière
weep hole	Sickerschlitz	barbacane
internal friction coefficient	Beiwert der inneren Reibung	coefficient de frottement interne
cohesive coefficient	Kohäsionsbeiwert	coefficient de cohésion
to pick someone's brains, (to extract information from someone)	jemandes Kenntnisse ausbeuten	exploiter les connaissances de quelqu'un
to benefit no end, (to benefit greatly)	in hohem Maße Nutzen ziehen	profiter indéfiniment
to force someone's hand, (to give someone no option)	jemandem keine freie Hand lassen	pas laisser la main libre à quelqu'un
to cast one's eye over something, (to look at something quickly)	etwas flüchtig ansehen	regarder rapidement
planking	Verschalung, Ausbohlung	revêtement
strutting	(Quer-)Aussteifung	entretoisage
pit	(Bau-)Grube	fouille
rectangular	rechteckig	rectangulaire
pressure gauge	Druckmesser	manomètre
to back up, (to support)	unterstützen, sichern	étayer

6. PLUMBING AND HOT WATER INSTALLATIONS

Both theory and practice have an equally important role to play in the planning of plumbing and hot water installations, and they should be suitably blended together to achieve the best possible results. Some forethought on the part of the designer in setting out the system and on the part of the plumber or heating engineer in executing the actual work can so often minimise the cost, as well as ensure greater success.
Many of the traditional basic materials, such as copper, cast iron, galvanised mild steel, bronze, gun metal, brass and zinc, used by the plumber in the past, are still much in evidence, but lead, the original material with which plumbers were always closely associated, is sometimes replaced by other less expensive and more readily available materials. In keeping with most other sections of the building industry, the present era of plastics has naturally had its impact on these installations as well. Plastics in many forms, e. g. polythene, or polyethelene (pe) and polyvinylchloride (pvc) are being used with some advantage in cisterns, rainwater goods, tubing and fittings and parts of taps and valves, to name just a few items. Fibreglass too has proved to be a useful substance, particularly in the manufacture of cisterns.

Largely due to the relatively high cost of copper, both phosphated and stainless steel could well present a challenge to copper for internal tubing. The gauge of most of the copper used for tubing and cylinders has inevitably become thinner over the years, and this has somewhat reduced its advantages. Plumbers are now experiencing more difficulty in producing smooth made bends and sets with a bending machine or a spring, because there is a tendancy for thin walled copper tubing to ruck. Even so, it is still regarded by many as the most suitable material for general use. The attractiveness of precoloured and plastic covered steel tubing and their competitive cost in relation to copper might encourage increased demand for them in time to come.

Theoretically, plumbing in Britain is divided into four distinct sections, namely rainwater installation, sanitary installation, cold water supply and sheet metal roof coverings. Rainwater installation includes everything necessary to convey storm water from a roof to the stormwater drain. In the case of a pitched roof, this is generally achieved by means of eaves gutters, or chuting as it is also known, and external downpipes. With a flat roof it is normally carried out through roof outlets and internal downpipes, but can also be effected by means of rainwater heads and external downpipes.

Sanitary installation comprises, (a) all the pipework and fittings necessary to dispose of the waste from sanitary fittings into the foul drain, (b) the sanitary fittings themselves, such as sinks, basins, baths, showers urinals, bidets, and water closets, or w. cs. as they are more often called, and (c) the overflow pipes and fittings required to convey surplus water to the outside of a building.

In large buildings, like hospitals, flats and office blocks, it is usual to arrange the sanitary fittings of one storey directly above those of the storey beneath. In this way, one soil and ventilation pipe is normally sufficient to serve all the sanitary fitting grouped together. In such cases, it is often necessary to incorporate antisiphonage pipes on all except the top storey. They form a vented connection between the traps of the sanitary fittings and the soil and ventilation pipe and prevent the water seals of the traps being drawn out by possible suction. This suction is caused by a fluctation in pressure in the soil pipe, normally after a w.c. has been flushed out.

Cold water supply consists of all the pipework and fittings required to convey water from the stop cock on the incoming water main, either direct or via a storage cistern, to the sanitary fittings, cylinder, and any independent draw-off taps. Accurate setting out and sizing of pipe runs are essential to ensure an adequate water supply at the correct pressure. Where practical, the shortest possible routes should be adopted, thus reducing pipework and fittings to the minimum, and as a result, decreasing the resistance to the free flow of water. Resistance can also be decreased a little by using such fittings as easy bends, pitcher tees and swept tees.

The supply of cold water to a sink, cistern and draw off tap, is normally taken direct from the high pressure rising main, and this is often, but not always, the case with other sanitary fittings as well. Where sanitary fittings or a cylinder are fed from a storage cistern, the supply is generally under low pressure. In some isolated country dwellings where no high pressure water main is available, a large capacity storage cistern is essential to guarantee a reasonable water supply over a period. In such a case, the source of water is usually a well, from which it is pumped to the storage cistern.

Sheet metal roof covering is already becoming a less significant part of plumbing, mainly because it is too expensive to use lead, copper or zinc for large flat roof areas. Tar or bituminous felt, asphalt and plastics have thus become more popular covering materials. Isolated metal coverings in connection with roofing, such as flashings, gutters and soakers are, however, still being executed by plumbers in most cases.

Hot water installation comprises (a) all the pipework and fittings necessary to convey hot water from the cylinder to those sanitary fittings which require a hot feed, (b) the primary flow and return pipes including fittings, also known as the main circuit, which connect the boiler and cylinder, (c) the ventilation pipe including fittings from the cylinder up to the storage cistern, (d) the cylinder and, (e) the boiler, whenever it is employed primarily to provide hot water. It is impossible to draw a distinct line between hot water and heating installations where the two are working in conjunction with each other, but the chapter on heating should help to clarify this point.

The remarks made with regard to short pipe runs and easy bends etc. in cold water supply apply equally to hot water installation. Accurate setting out and sizing of pipe runs is even more critical here than in cold water supply, since the conservation of heat is a further important factor to be considered. The hot water pipes and cylinder should always be lagged, where this is practical, so as to retain as much heat in the system as possible. Lagging of cold water pipes and cisterns is also necessary where they are exposed to extreme cold, to prevent the freezing of the water and its rather disastrous consequences.

Any economy in the use of valves in either cold or hot water installations often proves to be a false economy, particularly when inevitable repairs or alterations must be carried out at some future date. To be able to completely isolate a sanitary fitting or section of pipework by means of a valve during such a time, and so minimise any disruption to the system, more than justifies the initial extra cost of the valve. In fact, if expense is no object, it is advisable to provide isolating valves close to all sanitary fittings on both the cold and hot feeds.

Copper tubing generally can either be provided with screwed compression, welded or soldered capillary joints on couplings, elbows, bends, reducers, tees and other fittings. For the jointing of large diameter tubing, silver brazing is sometimes adopted. Each type has its advantage and disadvantage, depending upon whether practicability or appearance is the desired feature. Compression joints enable sections of tubing to be easily removed for repairs and alterations, but they are rather unsightly. On the other hand, welded, brazed and capillary joints are unobtrusive, but they make repair and alteration work a little more difficult, although not impossible.

The basic principles governing plumbing and hot water installation have seemingly altered very little over a long period of time, and it is only in the use of new materials that any radical changes have taken place. The role of the plumber, however, has drastically changed. Many years ago, his trade included leadwork of every description, even including lining coffins, heating and ventilation and drainlaying.

Vocabulary	Wörterverzeichnis	Glossaire
plumber	Klempner	plombier
plumbing	Klempner- und Installationsarbeiten	travaux de plomberie et d'installation d'appareils sanitaires
to set out	Arbeit vorbereiten und anlegen	préparation du travail
heating engineer	Heizungstechniker	technicien en chaufage
cast iron	Gußeisen	fonte
galvanised	verzinkt	galvanisé

mild steel	Flußstahl	acier doux
gun metal	Bronze	bronze
brass	Messing	laiton
lead	Blei	plomb
plastics	Kunststoff	plastique
polyethylene, polythene, pe	Polyäthylen, PE	polythène
polyvinylchloride, pvc	Polyvinylchlorid, PVC	pvc
cistern	Wasserbehälter, Wassertank	réservoir
rainwater goods	Erzeugnisse für Regenwasserableitung	tuyaux et raccords d'eaux pluviales
tubing	Rohrleitung	conduite
fitting	Formstück für Rohrleitungen	raccord
tap	Hahn	robinet
valve	Ventil	soupape
fibreglass, glassfibre	Faserglas, glasfaserverstärkter Kunststoff	fiberglas (fibre de verre)
stainless	rostfrei, nicht rostend	inoxydable
internal tubing	Rohrleitungen innerhalb des Bauwerks	canalisations intérieures
gauge	Wanddicke, Rohrwanddicke	épaisseur de la paroi
cylinder	zylindrischer Warmwasserbehälter (durch eingetauchtes Heizelement beheizt)	chauffe-eau
storey	Stockwerk, Etage	étage
soil pipe, soil stack	Abwasserfallrohr	tuyau vertical des eaux usées
bend	Bogen	coude
set	Versatzstück, Doppelbogen	double coude
spring	(Schrauben-)Feder	ressort helicoïdal
to ruck	Falten bzw. Knicke bilden	rider
rainwater installation	Regenwasser-Ableitung	canalisations d'eaux pluviales
sanitary installation	Schmutzwasser-Abführung	évacuation d'eaux usées
water supply	Trinkwasserversorgung	alimentation en eau potable
sheet metal	Blech	tôle
roof covering	Dachdeckung	couverture
stormwater	Regenwasser	eau pluviale
drain	Abflußleitung	drain
eaves gutter, chuting	Dachrinne	chéneau
downpipe, stack pipe	Fallrohr	descente
outlet	Abflußöffnung	ouverture
rainwater head	Regenwasserfang, Rinnenkasten	cuvette
pipework	Rohrleitung	tuyauterie
foul drain	Schmutzwasserableitung	canalisation d'eau usée
sink	Spülstein, Spülbecken	évier
basin	Handwaschbecken	lavabo
bath	Badewanne	baignoire
shower	Dusche	douche
flats, block of flats	Mietwohnungen, Wohnhaus mit mehreren Etagenwohnungen	appartements loués, immeuble de rapport
office block	Bürohaus	bâtiment administratif
antisiphonage pipe	Be- und Entlüftungsrohr für Geruchsverschluß, Sekundärentlüftungsrohr	reniflard de siphon
vented	belüftet	ventilé
trap	Geruchsverschluß	siphon
water seal	Sperre durch Wasser	obturation hydraulique

suction	Sog, Ansaugen	aspiration
fluctuation	Schwanken, periodische Änderung	fluctuation
to flush out	ausspülen	chasser
stop cock, stop tap	Absperrventil	soupape d'arrêt
water main	Wasserversorgungsleitung	canalisation principale en eau potable
storage cistern, feed tank	Vorratsbehälter, Vorratstank	réservoir d'eau
independent draw-off tap	einzelnes Auslaufventil, Zapfventil	prise d'eau
sizing	Bemessung	dimensionnement
easy bend, long radius bend	sanfter Bogen	coude aplati
pitcher tee, angle tee	45° Abzweig	branchement 45°
swept tee, sweep tee	Bogenabzweig	raccord courbé
rising main	Steigeleitung	tuyau ascendent
well	Brunnen	puits
felt	Dachpappe	carton feutre
flashing	Verwahrung, Blecheinfassung	bande
soaker	Kapp- bzw. Kehlblech	noquet
hot feed	Warmwasser-Speiseleitung	canalisation d'eau chaude
flow and return pipes	Zu- und Rücklaufrohre	canalisations de distribution et de retour
main circuit	Haupt(kreis)leitung	canalisation principale
boiler	Heizkessel	chaudière
pipe run	Rohrstrang	tuyauterie
to lag, to insulate	gegen Wärmeverlust dämmen, mit Dämmstoff ummanteln	isolation thermique
to isolate, to cut off, to separate	sperren, absondern, trennen	séparer
isolating valve, stop tap, stop cock	Absperrventil	soupape de fermeture
screwed compression joint	Quetschungsverschraubung	raccord à vis
to weld	schweißen	souder
to solder	weichlöten	souder à l'étain
capillary joint	haarfeine Fuge	joint capillaire
coupling	Rohrverbinder, Muffe	manchon
elbow	Winkel, Kniestück	raccord coudé
reducer	Reduzierstück, Reduktionsnippel	réducteur
tee	T-Stück	té double

Dialogue

Robert: It's well known, of course, that there is a growing tendency with all plumbing installations for plastics to take over the role of metals. Do you think this means the eventual disappearance of metals from the scene?

Duncan: Not entirely. I do realise, however, that with further research and developement, plastics will soon be suitable for practically anything, but alloys are also improving, Robert, as you know. In the long run, I believe cost will be the deciding factor.

Robert: So far plastics have had far less impact on hot water installations, but once the drawback of high temperature expansion has been overcome, I'm sure it will prove successful in these too. Prejudice against plastics, as with any other relatively new material on the market, will fade in time. Once it is fully accepted as a basic material, and craftsmen become accustomed to it and more experienced in its use, I'm convinced that it will largely supersede metals.

Duncan: I am rather interested to know if the German plumber, for example, is required to lay sheet metal roof covering, as the plumber in Britain still is on occasions.

Robert: No Duncan, that type of work is not his responsibility. It is executed by the roofer, who normally carries out the rainwater installation as well. In Hamburg, in fact, he even lays the stormwater drain directly associated with the rainwater downpipe.

Duncan: As we are on this particular subject of demarkation, it may be of some interest that plumbers and heating engineers in Britain sometimes fulfil the same purpose, to a limited degree, in hot water installations. Whenever heating and a hot water supply are installed in conjunction with each other, it is often the heating engineer's responsibility to install the hot water feeds up to the control valves close to the sanitary fittings. The short pipe connections, or tails as they are often known, from the valves to the sanitary fittings, are nearly always installed by the plumber. In cases where only a hot water supply is installed, the plumber carries out the entire work. Is this the procedure on the Continent?

Robert: Well Duncan, I can only really speak for Germany where it's quite different. Here, hot water supply and heating are always carried out by heating engineers. I think this is generally true in other Continental countries too.

Duncan: It might be as well at this point to mention some of the differing features of Continental and British plumbing.

Robert: Yes, I'm sure this would be of particular interest to many readers. Again, I can only draw accurate comparisons between German and British practice, with which I am rather more familiar. Probably, the main difference in the two systems is that we don't use cold water cisterns, as you do in Britain.

Duncan: What then replaces our cistern in your country?

Robert: Nothing at all, really. You see, all our fittings are connected directly to the incoming cold water supply off the mains.

Duncan: Isn't there a danger, in some cases, of soiled water being drawn back into the mains when they are empty during repairs?

Robert: That would seem likely, but to prevent this, an anti-suction valve is inserted at the highest point on the rising main inside the building, and incidentally, the fittings are actually connected to the dropping main below this valve.

Duncan: This system is, of course, very much cheaper than ours, but I feel the provision of a cistern provides a small emergency supply of water when the cold supply is cut off.

Robert: Yes, I'm rather inclined to agree. There is one particularly bad practice in Germany, which is rarely employed in Britain, I'm pleased to see, and that is the sinking of pipes into walls which are later plastered.

Duncan: Years ago, they often did the self same thing here, and the amount of trouble it caused when a leakage occurred soon made architects think again. Another thing which I find rather odd with German plumbing, is the way in which overflow pipes are allowed to discharge into baths or other sanitary fittings.

Robert: This I understand, is not generally allowed in Britain.

Duncan: You're right, because most water authorities are always anxious to prevent the waste of water. Where the overflow discharges to the outside of a building, any waste of water can usually be seen and often heard too.

Robert: I think it's high time that we clarified the expression "boiler", if you're agreeable.

Duncan: A good idea Robert, since it has widely different meanings in our two countries. In Britain, for example, it refers to the heat transfer, which is the means of burning the fuel to create the necessary heat output. Your term for this, I imagine, would be "Heizkessel", in the widest possible sense.

Robert: That's quite right, but we also use the English word "boiler", which refers to an independent electric hot water heater which is not insulated. The same item which is insulated and enclosed in an enamelled sheet metal casing is known as a "Speicher".

Duncan: Moving on to another subject, may I ask you which form of energy you prefer to produce hot water?

Robert: That rather depends on the cost of both the energy and the installation of the necessary apparatus to utilise it. On carefully weighing up all the factors involved, I've come to the conclusion that electicity is preferable to oil, gas or solid fuel. The main advantages, I feel, of an electrical system are that it can usually be installated pretty economically and it's clean and simple in operation. Take, for example, a system using insulated and thermostatically controlled single-point water heaters, each one being situated close to the sanitary fitting it serves. Hot water is always at hand in an instant and installation costs are not excessive.

Duncan: I agree, Robert, that independent heaters are effective, but I suspect that this particular system is less economical than other comparable electrical systems. Don't you think that the use of a single-insulated and thermostatically controlled instantaneous multi-point water heater, such as the "Durchlauferhitzer" which I have often heard mentioned in Germany, would provide a cheaper method.

Robert: Yes, you may well be right. Both systems are, in fact, widely used here, but I would point out that the Durchlauferhitzer, requires a high intensity of current in order to provide sufficient instant heat, and permission to use a heavy supply cable must be obtained from the electricity authority. For either system to be really effective, short pipe runs are essential , as you can appreciate, to avoid undue loss of heat. By the way, is it allowed in Britain to connect the cold feed of a water heater direct to the authoritys' water main?

Duncan: In the case of a pressure type heater, which has no permanently open outlet, it is not normally permitted. It should be connected to a cold feed tank, which is a cistern controlled by a ball valve.

Robert: In your view, is there any justification in the allegation by some people, that electrical hot water appliances are a danger to life?

Duncan: I hardly think so in the vast majority of cases, expecially if the appliance has been produced by an approved manufacturer and properly installed by a qualified electrician. Naturally, the same principles apply to gas water heaters, or geysers as the older types were formerly called. Talking of these gas appliances, it is important, of course, that they should always incorporate a flame failure device, to avoid any dangerous gas leakage, especially where gas is odourless and thus more difficult to detect. It is also essential that a separate flue is provided for the escape of fumes from the outlet of a gas appliance.

Robert: Before we wind up our somewhat varied discussion, Duncan, I would hasten to add that one ought never to overlook the importance of suitable mains water. For example, thorough treatment of hard water long before it enters the system is necessary, to avoid the bore of the pipes becoming gradually choked with deposits, resulting from chemical action of the water on the walls of the pipes.

Duncan: I wholeheartily agree that prevention is better than cure, to use a well known proverb, and I can't think of a more appropriate way in which to draw this subject to a close.

Vocabulary	Wörterverzeichnis	Glossaire
alloy	Legierung	alliage
roofer	Dachdecker	couvreur
stormwater drain	Regenwasser-Abflußleitung	canalistion d'eaux pluviales
demarkation	Abgrenzung	démarcation
control valve	Steuerventil	soupape de commande
tail, short pipe connection	Verbindungsrohr	tube de raccord
heating engineer	Heizungsbauer, Heizungsmonteur	technicien en chauffage
cold water supply	Trinkwasserversorgung	alimentation en eau potable
main	Hauptleitung	conduite principale
soiled	verschmutzt, verunreinigt	souillé
anti-suction valve	Rückstauventil	clapet de non retour
emergency supply	Notversorgung	alimentation de secours
sinking into walls	Unterputzverlegung	pose sous enduit
to plaster	mit Innenputz versehen, putzen	enduire
authority	Behörde, Amt	administration publique
heat transfer	Heizkessel	chaudière de chauffage

independent hot water heater, (not insulated)	Boiler	chauffe-eau non isolé
independent hot water heater (insulated)	Speicher	chauffe-eau isolé
insulated single point water heater	Heißwasserspeicher mit einem Zapfventil	chauffe-eau isolé à robinet unique
instantaneous water heater	Durchlauferhitzer	chauffe-eau instantané
instant heat	sofortige Wärme	chaleur instantanée
cold feed	Frischwasserzulauf	alimentation en eau froide
authority's water main	öffentliche Wasserversorgungsleitung	canalisations publiques
pressure type heater	Druckspeicher	chauffe-eau à accumulation et pression
ball valve	Schwimmerventil	soupape commandée par flotteur
hot water appliance	Heißwassergerät	chauffe-eau
gas water heater, geyser	Gas-Wasserbereiter	chauffe-eau à gaz
flame failure device	Gaswächter	interrupteur de sûreté
flue	Schornsteinzug	conduit de fumée vertical
to wind up	zu Ende bringen	achever
mains water	Wasser der öffentlichen Wasserversorgung	eau de la ville
treatment	Aufbereitung	préparation
bore	lichte Weite	diamètre intérieur
choked	verstopft	colmaté
deposit	Ablagerung	sédiments

"I thought you said that changing a washer was the easiest thing in the world."
(Reproduced by permission. BAUER, Esslingen)

7. HEATING AND VENTILATING INSTALLATIONS

The essence of really efficient heating and ventilating installations lies fundamentally in a properly thought out design, although the careful choice of the most suitable materials for each particular system and a high standard of workmanship do have some effect on the overall success of any scheme. The design is governed, to a large extent, by a number of closely associated factors. These are the required inside air temperatures, known as design temperatures, the heat losses and the desired frequency of air changes for each room or area of a building to be served.

The design temperatures depend primarily on the extent of space heating required – in other words on whether full or just background heating is called for. Quite often only background heating, providing an air temperature around 18° Celsius, is necessary in such areas as halls, stairways and stores. In areas such as residential rooms, offices and hospital wards, where full heating is usually required, the air temperature is normally between 20° and 25° Celsius. To maintain the design temperatures throughout a building, taking into consideration the maximum heat losses, the boiler or other means of heat transfer must comfortably provide the necessary output. This output, or rating as it is also known, is expressed in kcal/h.

Heat losses, or heat transmission, from rooms etc. fluctuate with the varying differences between the inside and outside air temperatures. They are at their maximum when the outside temperature is at its lowest, thus causing the warmer inside air to flow rapidly towards the far colder outside air.

The particular materials used in the construction of a building have a considerable bearing on the extent of heat losses. Some materials provide better barriers than others to the transmission of heat. For example, a brick wall plastered internally and rendered externally is a more effective barrier than a plain brick wall of equal thickness. Similarly, a boarded roof covering offers greater thermal insulation than an unboarded one. There are other lesser, but nevertheless accountable factors which affect the amount of heat lost from a building. Wall cracks, constructional gaps due to illfitting windows and doors, air ventilators and flues are just a few. In order to make a reasonably accurate calculation of heat transmission possible, each material is given a value, which can be readily found in standard tables.

This value is, in fact, the coefficient of heat transmission, or thermal transmittance of a material. It is expressed as the number of kcals transmitted through a square metre of construction 1 metre thick for one degree Celsius difference in temperature between the inside and outside surface. Applying this value for the actual thicknesses of the various materials in the construction of the walls, floors, roofs etc., the thermal insulation value of each part of the construction can be calculated approximately. With this, plus the allowances for air transmission, the calculation for the heat requirement can be determined.

The next step is to select the type of system capable of producing a heat output to match this requirement and to suit the general layout of the building. The choice of heating systems is wide, both in the means of producing heat and in the type of installation. Electricity, gas, oil or solid fuel provide the alternative means of producing the heat requirement, whilst hot water, steam, warm air and electric heating elements are the most common media through which the heat is actually emitted at the necessary locations.

At the present time, many space heating systems, especially those for domestic purposes, utilise hot water as the medium, and this is distributed through pipes to the radiators or other forms of heat emission. The means of heat transfer is usually an independent oil-, gas- or coal-fired boiler situated in the lowest part of the building. There are several differing forms of hot water heating system, and broadly speaking, these fall into two main categories, namely low and high pressure. In the low pressure, or open expansion tank system, the water is open to the atmosphere, whereas the high pressure system is hermetically sealed.

The low pressure system is particularly suitable for most dwellings and other small buildings, since it is unlikely to suffer damage through neglect, nor is there any likelihood of an explosion. Furthermore, it requires very little attention to function successfully and is not prone to failure. However, it has disadvantages too, especially in the more extensive gravity circulation systems, in that large radiator heating surfaces and distribution pipes are necessary, because the system operates with a maximum temperature of around 100° Celsius and has a temperature range of about 12° Celcius.

Water circulation in a low pressure system can be improved appreciably, resulting in better heat distribution, with the inclusion of an accelerator, thus giving what is known as a forced circulation. In recent years, in fact, the production of many efficient, economical and silent accelerators, capable of circulating a small amount of water against comparatively high-friction heads, has made small-bore and even mini-, or micro-bore central heating installations both possible and practical for domestic use.

Basically, a low pressure system usually comprises a boiler, water feed and expansion tank, flow and return pipes forming the circuit, or main as it is also termed, and radiators connected to the main by two short branches. Where an oil-fired boiler is adopted, an oil storage tank and feed pipe also form part of the installation. When a large boiler is coal- or coke-fired and situated in an independent boiler house, either a hopper or worm and conveyor feed is normally necessary, depending on whether the boiler is fed from above or below.

The arrangement of the circulating pipes and radiators vary from one type of installation to another to give either a single-pipe or two-pipe system. With the single pipe system, all radiators are served by one common main via branches, the feed branch coming off the top of the main and the return branch usually being tapped into the side of the main. The two-pipe system employs two separate mains, one for the flow of hot water from the boiler to the radiators and the other for the return of cooler water back to the boiler from the radiators. Within these two arrangements of pipework and radiators just described, there are further divisions, namely the overhead, or drop, and up-feed systems, depending on whether the radiators are fed from above or below.

"I can never understand why my wife doesn't appreciate regular maintenance."
(Reproduced by permission. BAUER, Esslingen)

The single-pipe system, having economy as its main virtue, has the marked disadvantage of necessitating extremely careful proportioning of radiator sizes to ensure uniform temperatures throughout the circuit. Otherwise those radiators nearest the boiler would benefit from a high temperature, whilst those furthest away would suffer as a result. To a certain extent, the system can be evenly balanced by adjusting the radiator lock shield valves on the return branches, and the heat emission of each radiator can be easily controlled by regulating the radiator valves on the flow branches. This particular system often proves sluggish if gravity circulation alone is relied upon, but in a forced circulation, this fault is largely dispelled.

The high pressure is often more suitable than the low pressure system for many large commercial, industrial and public buildings, mainly because its working and maintenance costs are comparatively low and its heat carrying capacity is very much higher. In such a system, water usually circulates at a temperature around 180° Celsius. Owing to the increased pressure, the pipes and fittings must be of heavy gauge, but they are invariably of relatively smaller bore. The main components of this system differ very little from those of the low pressure system, except that an expansion tube usually replaces the open expansion tank, and radiators are replaced, where necessary, by heating coils forming part of the main, high level radiant panels, overhead strip heating or other forms of heat concentration. Coils are sometimes welded to sheet steel plates to provide increased heating surfaces, or are embedded in concrete floors or enclosed in the ceiling or wall construction. High pressure hot water systems are often preferred to their steam counterpart, principally because they are cheaper to install.

Whatever form of hot water heating system is adopted, there are certain fundamental principles to be observed to ensure its proper function. The main ones are, that the pipes should be properly proportioned to carry an adequate supply of hot water to all the radiators or alternative heating surfaces, and that all radiators, where employed, should be provided with an air cock at their highest point to release any trapped air. It is also advisable that the flow pipe should be inclined upwards towards each radiator to allow trapped air to reach the top of the radiator, and the return pipe inclined downwards away from each radiator. The lagging of pipes is vital where they are employed solely for heat distribuion and not partly as a heating surface as well.

Besides hot water heating, there is a wide choice of heating systems to suit the many differing demands and circumstances of both industrial and domestic users. The most prominent alternatives are steam, warm air, radiant and underfloor electric heating. Gas radiant heating is usually employed for industrial purposes, whereas its electric counterpart is more commonly found in dwellings.

Steam heating has many basic features in common with hot water heating, but is sometimes more suitable for extensive schemes. In district heating, for example, steam can be employed with heat exchangers to provide hot water heating for a whole area of domestic buildings. In such schemes the steam is often provided from waste heat of industrial plant, like an electric or nuclear power station or a steelworks. Two of the major problems to be overcome in a steam heating system is the expansion and contraction of the mains and condensation in the mains. To counteract the first of these, the pipes in the ducts are usually supported on special brackets often fitted with rollers to permit free movement, and expansion pits are provided at intervals along the duct, in which the pipes are bent around to take up the expansion or contraction. To remove the water formed by condensation when heat is given up by the steam, the pipes should be laid to fall in the direction of the flow of steam and drain pockets provided at regular intervals. Any water remaining in the main will cause sharp knocking, commonly known as water-hammer.

Warm air heating really began in the times of the Romans, and the system they employed, which has since been revived and notably improved, was that which we now broadly describe as the indirect system. Warmed air is passed through ducts incorporated in the structure of the building, which is thus heated. Alternatively, there is the more commonly adopted direct system, in which the warmed air is actually passed into the spaces to be heated, either by natural convection currents or mechanical propulsion.

The direct system has many forms, but in the more general sense, the principal two are the local and remote systems. The former is so called, because the spaces to be heated are each served by independent air heaters actually positioned, either at low or high level, in the spaces themselves. These heaters are self-contained units, complete with burner, combustion chamber, centifugal fan, air inlets, warm air discharge grilles, flue outlet and controls. The remote, or plenum system, comprises a central air heater, main and branch ductwork through which the warmed air is distributed throughout the building and the fresh air drawn in, an air inlet and warm air outlet grilles. The air heater in this system has much in common with those for the indirect system, except that it is normally larger and has a greater output to cater for a wider demand.

Most air heaters in the systems mentioned derive their heat directly from gas, oil or electricity. In heaters burning gas or oil, the resulting hot gases pass through a combustion chamber and warm the air circulating around the outside of the chamber. The gases are subsequently removed through an outlet into the flue. Where electricity is utilised, the air to be warmed is passed over an electric heating element.

The advantages that a warm air heating system has over its rivals are that it produces fairly rapid results, and in the case of a ducted system, it can be used in summer time with the heat off purely as a cooling system, which is particularly desirable in industrial buildings. Its main disadvantages, on the other hand, are that in the ducted system much space is often required for the installation and it is rather costly, and that it often needs some other form of local heat, such as electric fires, to supplement its heat output. It is important to finely control the velocity of flow of air in any mechanically propelled system, otherwise uncomfortable draughts will be created with too high a velocity, or inadequate air changes will take place with too low a velocity.

When dealing with some types of central heating system, especially those adopting solid fuel, gas or oil burning boilers as the heat transfer, the design of chimneys must be considered as part of the overall design of the installation. Balanced flues used in connection with some gas burning boilers do, however, eliminate the need for chimney flues.

In the construction of a building, the location, size and height of chimneys are important factors. In order that chimneys may draw efficiently, each boiler outlet, where more than one boiler exists, should have its own separate flue, and where possible, they should be grouped together to minimise the cost and improve the efficiency of the chimneys. To avoid possible down draughts passing directly into the boilers, flues are constructed with smooth gradual bends in their lengths. Flues are normally lined with special sectional flue liners, often capped at the top with terminal blocks, and this lining assists the speedy passage of smoke and fumes quite considerably. A chimney stack should be built to a height of at least 1 metre above the top of the roof, so that the smoke and fumes are carried away clear of the building.

Under-floor electric heating has become more popular over the years, especially in such public buildings as schools and libraries etc., where large unobstructed floor and wall areas are conducive to the efficient running of the establishment. The two alternative methods are to embed the heating cable coils in the floor screed or to bury small ducts therein and feed the cables through them. The former method is considerably cheaper to install, but the latter has the important advantage of providing easy access at any time for repairs or replacements.

Ventilating a building, broadly speaking, merely means removing the vitiated air from inside and replacing it by fresh air from outside. In the average dwelling or small building, this may be effected satisfactorily through a natural process – that is by means of holed, louvred or "hit and miss" ventilators in the external walls, opening lights of windows, vented fanlights over doors etc. The movement of air is often assisted, even in very austere systems, by fans built into external wall openings or incorporated in windows, especially in kitchens.

For larger buildings with special needs, far more elaborate systems have to be adopted and these often go hand in hand with ducted warm air heating systems. Two distinct methods of ventilation are commonly practised, namely the input system, whereby fresh air is drawn in through a duct by a fan and vitiated air is removed by natural means, and the extract system, where the process is reversed. A balanced ventilation system, that is a combination of both an input, or plenum, and an extract system, is the most desirable, as all air movement can be controlled.

In those public buildings, like assembly halls, cinemas and concert halls, where large concentrations of people must be catered for, and hence frequent air changes are necessary, air conditioning accompanies ventilation, and sometimes warm air heating too. Air conditioning can be simply defined as the treatment of incoming air, so as to control its cleanliness, humidity, temperature and distribution to comply with the requirements of the occupants of the particular building. Air washers constitute an essential component of any full air conditioning system and these normally consist of a chamber at the entrance of the duct, in which a fine mist is produced from water pumped under pressure through jets. Air passing through the chamber is cleaned, humidified and cooled in one process. There are a number of additional processes, such as air filtration, etc., which often form part of complex systems in special circumstances.

The conservation of heat is an important factor in any type of heating system, for it enables real economies to be made in the installation. This can be made effective in the actual construction of the building with the use of materials having good thermal insulation properties. The construction of external walls in lightweight thermal concrete blocks, filling the spaces between ceiling joists with a glassfibre quilt and the use of double glazing are just a few examples of methods of conserving heat. It is common sense to insulate a building as far as is practical, not only to make a heating system more effective, but also to assist a ventilating and air conditioning system in hot weather.

Vocabulary	Wörterverzeichnis	Glossaire
design	Entwurf, Bemessung	étude
design temperature	Bemessungstemperatur	température de calcul
heat loss	Wärmeverlust	perte de chaleur
space heating	Raumheizung	chauffage de locaux
background heating	Grundheizung	chauffage de base
stairway	Treppen	escalier
store	Lager	magasin
residential room	Wohnraum	chambre d'habitation
office	Büro	bureau
hospital ward	Krankenzimmer	chambre des malades
boiler	Heizkessel	chaudière
rating	Nennleistung	rendement nominal
heat transmission, thermal transmittance	Wärmedurchgang	transmission thermique
brick wall	Ziegelmauer	mur en briques
to plaster	mit Innenputz versehen	enduire à l'intérieur
to render	mit Außenputz versehen	enduire à l'extérieur
plain wall	ungeputzte Wand	paroi nue
to board	mit Holzverschalung versehen	revêtir en bois
boarded roof covering	hölzerne Dachverschalung	couverture en bois
thermal insulation	Wärmedämmung	isolation thermique
crack	Riß	fissure
gap	Spalte	fente
illfitting	schlecht passend	convenant mal
air ventilator	Öffnung für Lüftung	ouverture d'aération
flue	Schornsteinzug, Rauchabzugskanal	conduit de fumée vertical
standard table	Normentafel, Normentabelle	tableau de normes
thermal transmittance coefficient	Wärmeleitzahl	coefficient de conductibilité
allowance	Zulage	supplément
air transmission	Wärmeübertragung an die Luft	transmission de chaleur à l'air
heat requirement	Wärmebedarf	quantité de chaleur
solid fuel	fester Brennstoff	combustible solide
radiator	Heizkörper	radiateur
independent boiler	freistehender Heizkessel	chaudière indépendante
low pressure	Niederdruck	basse pression
expansion tank	Ausdehnungsgefäß	vase d'expansion
open expansion tank system	"Offenes System"	système ouvert
hermetically sealed	luftdicht verschlossen	étanche à l'air
distribution pipe	Verteilungsrohr	tuyau de distribution
temperature range	Temperaturbereich	limits de temperature
accelerator, circulation pump	Umlaufpumpe	pompe de circulation
head	Druckhöhe	hauteur de refoulement
friction	Reibung	frottement
bore	lichte Weite	diamètré intérieur
feed	Speiseleitung, Zuleitung	canalisation d'alimentation
flow pipe	Vorlaufrohr	tuyau de trop-plein
return pipe	Rücklaufrohr	tuyau de retour
circuit	Leitungskreis	circuit
main	Hauptleitung	conduite principale

branch	Zweigleitung, Anschlußleitung	conduite secondaire
storage tank	Lagertank, Lagerbehälter	réservoir
boiler house	Kesselhaus	chaufferie
worm feed	Beschickung durch Förderschnecke	alimentation par vis sans fin
conveyor feed	Förderband-Beschickung	alimentation par bande transporteuse
single-pipe system	Einrohrsystem	chaufferie à tube unique
overhead system, drop system	System mit oberer Verteilung	chaufferie collective descendante
up-feed system, rising system	System mit unterer Verteilung	chaufferie collective ascendante
proportioning	Dosierung	dosage
lock shield valve	Drosselventil mit Abdeckkappe, Behördenventil	papillon à capot
radiator valve	Heizkörperventil	soupape de radiateur
gravity circulation	Umlauf durch Schwerkraft	circulation par gravité
working cost, running cost	Betriebskosten	coût d'exploitation
maintenance cost	Unterhaltungskosten	coût d'entretien
heat carrying capacity	Wärmekapazität, Wärme(energie)-inhalt	capacité thermique
fitting	Formstück, Rohrverbindungsstück	raccord
heating coil	Heizschlange	serpentin de chauffage
radiant panel	Strahlungsheizkörper	panneau rayonnant
strip heating	Lamellen(grill)heizung	chauffage par radiateurs à ailettes
to lag	dämmen	isoler
district heating	Bezirks-Fernheizung	chauffage urbain
heat exchanger	Wärmeaustauscher	échangeur de chaleur
nuclear power station	(Atom-)Kernkraftwerk	centrale électrique nucléaire
duct	Kanal	gaine
bracket	Konsole, Halterung	console, appui
expansion pit	Ausdehnungsgrube	fosse d'expansion
drain pocket	Sumpf	puisard
convection current	Konvektionsströmung	courant de convection
mechanical propulsion	Zwangsförderung	air pulsé
level	Höhe	niveau
self-contained unit	selbstständige Einheit	unité indépendante
burner	Brenner	brûleur
combustion chamber	Verbrennungsraum	chambre de combustion
centrifugal fan, fan	Ventilator, Lüfter	ventilateur centrifuge
air inlet	Lufteinlaß	entrée d'air
air discharge grille	Luftaustrittsgitter	grille de sortie d'air
control	Bedienungsorgan	organe de commande
main and branch ductwork	Haupt- und Nebenkanäle	gaines principales et secondaires
electric fire	elektrischer Ofen	appareil de chauffage électrique
to control finely	feinregeln	régler avec précision
velocity of flow	Strömungsgeschwindigkeit	vitesse de courant
mechanically propelled system	Zwangsförderungssystem	système d'alimentation forcée
draught	Luftzug	tirage
air change	Luftaustausch	échange d'air
heat transfer	Wärmeübertragung	transmission de chaleur
chimney	Schornstein	cheminée
overall design	Gesamtplan, umfassender Entwurf	plan de ensemble
balanced flue	Strömungssicherungsanschluß, kombiniertes Abgas-Frischluft-System	conduit de fumée vertical équilibre

boiler outlet	Heizkesselabzug	sortie.de chaudière
efficiency	Wirkungsgrad, Wirksamkeit	rendement
to line	auskleiden, bekleiden	revêtir
liner	Auskleidungselement	élément de revêtement
sectional	geteilt (aus Abschnitten) zusammensetzbar	composable
terminal block	Schornsteinkopf(-Stein)	mitron, mitre
smoke	Rauch, Qualm	fumée
fume	Dampf, Rauch	vapeur, brume
chimney stack	Schornstein	cheminée domestique
cable coil	Kabelschlange	serpentin de câbles
screed	Estrich	chape
vitiated air	verbrauchte Luft	air vicié
"hit and miss" ventilator	regelbares Lüftungsgitter	grille d'aération réglable
opening light	Fensterlüftungsflügel	vantail à ouvrant
to vent	lüften	ventiler
fanlight	Oberlicht über Tür	fenêtre de dessus-de-porte
input system, plenum system	Überdruck-System	système à surpression
extract system	Unterdruck-System	système à vide
air conditioning	Klimatisierung	conditionnement d'air
humidity	Feuchtigkeit	humidité
jet	Düse	gicleur
thermal block	großformatiger Wärmedämmstein	élément préfabriqué réfractaíre

Dialogue

Duncan: When one thinks about heating, the mechanical aspects of the installation usually spring to mind, and other considerations are invariably brushed to one side as being rather unimportant.

Robert: What exactly are you driving at, Duncan?

Duncan: What I'm trying to put over is that the human body has to be considered in all heating systems.

Robert: I think I've cottoned on to your theme. In fact, I imagine you're drawing a parallel between the human body and a stove, both of which give out heat.

Duncan: Yes, Robert, you're right on the ball. To give a few simple facts, the human body emits approximately 100 kcals per hour,and in cold weather the surrounding temperature provided by a heating system should be sufficient to prevent a greater quantity being given out. In other words, a balance should be maintained to avoid discomfort.

Robert: This leads me to produce further statistics in this particular field. As most of us know, greater heat is necessary to provide a comfortable air temperature in a dry atmosphere than in a humid one. To be a little more specific, I should add that hygenic research has proved that on average we feel most comfortable at a temperature of between 18° und 20° Celsius in 40% to 50% of humidity, but when the temperature is as low as 16° Celcius, for example, the humidity should be just under 70%.

Duncan: Turning to one of the more usually discussed topics, I'd like your view on whether you think the thermal calculation of a building is an important factor to be specified.

Robert: I would go as far as to say that I regard it as an essential factor. The thermal insulation of an external wall, for example, must be given a minimum value to comply with the Government's requirements for each particular heat zone.

Duncan: This stipulation is for dwellings, isn't it?

Robert: Yes it is, and the beauty of it is that it provides the client with some indication of the quality of the structure, especially in prefabricated houses.

Duncan: In fact, it's a useful safeguard for the house hunter, which I regret to say is sadly lacking at the moment in Britain.

Robert: Perhaps we can go into the subject of thermal insulation a little deeper.

Duncan: By all means. Well, you start the ball rolling, Robert?

Robert: Certainly. The lining or infilling of walls with modern insulation materials, such as expanded polystyrene and pressed wood fibre, has a marked effect on heat conservation, as you know, and at long last I believe the building technologists have succeeded in driving this point home.

Duncan: This is true, but I think one must take this a step further and concentrate on building walls etc. entirely of insulation materials, rather than have this sandwich type construction using a combination of conventional and insulation materials.

Robert: Obviously, you have in mind homogeneous lightweight concrete having such aggregates as foamed slag or mica. Of course, this is already in use, and maybe the fully cast lightweight concrete house is only just around the corner.

Duncan: A very interesting and inseparable part of heating, I know, but do let's give some time to actual heating installations before we lead our readers into a general debate on building techniques.

Robert: Yes, you're right. Turning then to heating systems, it's odd to find that old ideas, which were often thought to be gone for ever, are sometimes revived – I mean, for example, warm air heating, which the Romans utilised, with some success, principally to heat their floors.

Duncan: Heated floor systems are very agreeable – I wouldn't dispute that – but on their own they are only suitable in mild climates. In cold climates the heat emission would have to be adequate to serve the whole building, and this would cause considerable discomfort to the feet where the heat is most concentrated.

Robert: Ideally, in cold climates, such a system should be augmented by a secondary means of heating, but this entails additional expense. For this very reason it is normally ruled out.

Duncan: Tell me, Robert, what do you see as the prime source of energy for our heating systems of the future?

Robert: Electricity without doubt, in my view.

Duncan: So you really think that gas, oil and solid fuel will fade out?

Robert: That's rather a loaded question. If you are only referring to heating systems in general, my answer is yes, but if your question is rather more vague, as I suspect it is, knowing you, then my answer is no. You see, industry can't do without the other sources of energy, and I don't see any dynamic change in industrial methods for some years to come.

Duncan: The fact that atomic energy will be transformed into electrical energy does reinforce your view about electricity, I must say. Furthermore, the distribution of electricity is so easily and neatly carried out. Thinking about it, I'm rather inclined to share your view.

Robert: Over the years we have moved far away from the open fireplace, which I must confess I have so often enjoyed in your lounge in England. Despite the warmth, comfort and atmosphere of this rather romantic feature, you do agree, I'm sure, that it is impracticable, and can only remain as a pure luxury.

Duncan: I won't argue that point, which I willingly concede.

Robert: In connection with small bore heating, you have often talked about an accelerator, which is surely a pump, or am I mistaken?

Duncan: It is a form of pump in fact, but more often termed an accelerator, or even a circulator. If you want to be absolutely understood, call it a circulating pump.

Robert: If I dare say it – how typically English. Anyway, joking apart, a circulating pump is essential in most small bore and in all mini-bore systems, so that a greater amount of hot water can be circulated.

Duncan: I would draw your attention to an important point here. A pump (excuse the abbreviation!) should be designed so as to offer little resistance to the flow of water, should it fail, otherwise a by-pass pipe is necessary to keep the system functioning at all.

Robert: To switch to something rather different, Duncan, would you care to make some comments about electric storage heaters and how well they have caught on in Britain?

Duncan: Why, of course. The simple answer to your second point is that they are quite popular here in Britain, mainly because they make use of "off-peak" electricity at the cheaper night tariff, and also because they are neatly constructed and often attractive. From the practical point of view, they are ideal for installing in old dwellings, since very little disturbance of the existing structure is necessary.

Robert: They're all valid points that you've made, and I would just like to add that the most modern forms are often fitted with fans and are fully automatic and controlled by electronic elements, thus requiring no attention whatsoever. These factors alone weigh quite heavily in their favour.

Duncan: Moving on to the next point, I believe I'm correct in saying that radiant heating is coming in on the Continent, especially, in industrial buildings.

Robert: Yes, this is so, and incidentally, what kind of future do you think it has?

Duncan: In industry I feel it has good prospects, mainly on the grounds of economy.

Robert: For what reason do you consider it's economical?

Duncan: Because radiant heaters only serve those parts of a building which really require heating – in the case of a factory, for example, the areas where the work is actually being carried out.

Robert: This is certainly true, and satisfactory results have been achieved in many factories by locating radiant heaters at high level – often just beneath the roof in fact.

Duncan: May I enlarge on this, Robert, by saying that from such heights these heaters are capable of effectively warming the hard floors below, from which heat is emitted to produce a comfortable working temperature. Before finally leaving radiant heating, I would mention two of its overall advantages – these are, that it provides instant heat and has a relatively low installation cost.

Robert: The advantages you've mentioned make this form of heating particularly practical for buildings like concert halls, sports pavillions, churches and the like, which are only used intermittently. As a matter of interest, I've seen it used with success to heat half-open cafe terraces, open air fair grounds and large marquees.

Duncan: Of course, it must be remembered that this system, like the underflow heating system discussed earlier, can seldom function adequately as the sole system for an entire building. In other words, toilet and storage areas etc. usually call for an independent supply of heat.

Robert: If you're agreeable, Duncan, I think heat reclamation is the next item worthy of discussion.

Duncan: Agreed, so why not begin, Robert, with those interesting facts which so impressed you about the supermarket near Düsseldorf.

Robert: Ah yes! In this particular supermarket much of the waste heat emitted from customers, fridges, cooling plant, lighting etc. was ducted away, compressed to raise its temperature, and then used to heat incoming fresh air for re-heating the building. Of course, such a system also served to ventilate the building.

Duncan: On the face of it, it seems a logical system and one which is bound to appeal to those endeavouring to cut heating costs.

Robert: One of the most striking facts that has emerged from research into such systems, is that as much as 80 % of waste heat from fluorescent lighting tubes can often be reclaimed.

Duncan: Referring back to that supermarket, I understand that very little extra heating was necessary, even in cold weather. Was that really so?

Robert: Yes indeed. In fact, the system was so successful, that the manager strongly advocated its use in all the other supermarkets in the chain.

Duncan: As you are well aware, there are many sources from which heat can be reclaimed – take sewage works and burning garbage, to name but two examples. To my mind, there's tremendous scope in this direction.

Robert: It has possibilities, I've no doubt, but I can't help thinking that it's a rather complicated system to regain such a comparatively small amount of energy. I wonder that the enormous radiant energy of the sun can't be utilised far more.

Duncan: Here again, I think the scope for research is enormous, and I can foresee the sun playing an important role in future heating schemes.

Robert: You may be right, but to bring us all back to earth, for the moment anyway, may I wind up this discussion with a point about the architect's position regarding heating and ventilation systems.

Duncan: Go right ahead, Robert.

Robert: With so many alternative systems these days, the architect needs a wide knowledge to advise his client on the most suitable system for his requirements.

Duncan: Yes, it's true his knowledge has widened, but in Britain such advice is normally the responsibility of the consultant on most contracts. However, since the architect's overall design of a building is effected to an extent by the heating and ventilating system, he must have some understanding of each system.

Robert: Doubtless we could chat on for some time, but I think it's best to call a halt and give our readers another breather.

Vocabulary	Wörterverzeichnis	Glossaire
to spring to mind, (to be recalled)	augenblicklich einfallen	avoir une idée subite
to brush to one side (to ignore)	"beiseiteschieben", nicht beachten	écarter
to drive at (to aim for)	hinzielen auf	en vue de
to put over (to explain)	ausdrücken, sagen	exprimer
to cotton on to one's theme (to understand an idea)	allmählich verstehen	y venir peu à peu
to be right on the ball (to be well informed)	"richtig liegen", rechthaben	avoir raison
stipulation	Vorschrift	prescription
the beauty of it (the advantage of it)	das Gute daran	l'avantage de
indication	Angabe	indication
client	Bauherr	maître de l'œuvre
safeguard	Schutz, Sicherung	protection
house hunter	Hausinteressent	client
expanded polystyrene	Polystyrol-Hartschaum	polystyrène expansé
wood fibre	Holzfaser	fibre de bois
to drive a point home (to make something perfectly clear)	eine Sache klarmachen, etwas verständlich machen	faire comprendre quelque chose
aggregate	Zuschlagstoff	granulat
foamed slag	Schaumschlacke, Hüttenbims	ponce de laitier
expanded mica	Blähglimmer	mica expansé
to be only just around the corner (to be close)	unmittelbar vor der Tür stehen	être juste à coté
heat emission	Wärmeabgabe	émission de chaleur
to rule out (to dismiss)	nicht in Betracht ziehen	ne pas prendre en considération
to fade out (to disappear)	(allmählich) verschwinden	disparaître peu à peu
loaded question	vielschichtige Frage	question complexe
open fireplace	offener Kamin	cheminée
storage heater	Speicherofen	appareil de chauffage à accumulation
"off-peak" electricity	Nachtstrom	courant de nuit
radiant heating	Strahlungsheizung	chauffage rayonnant
heat reclamation	Wärmerückgewinnung	récupération de chaleur
cooling plant	Kühlanlage	installation de réfrigération
lighting	Beleuchtung	éclairage
to duct away	in Kanälen wegleiten	évacuer dans des gaines
on the face of it (at a glance)	auf den ersten Blick	sur le coup
sewage works	Kläranlage	station d'épuration
burning garbage	Müllverbrennung	crémation des ordures
consultant	beratender Ingenieur	ingénieur-conseil

8. SURVEYING

Surveying in the broadest possible sense covers many forms, including geodetic, hydrographical, topographical, cadastral, mining and engineering. Geodetic surveying is carried out on extremely large areas of land, where the earth's spheroidal shape has to be taken into account. Such surveying forms the basis for less extensive surveys within these larger areas. Hydrographical surveying is employed solely on the sea and other expanses of water. For producing maps depicting the principal ground features of towns, villages, roadways, rivers, forests, and the like, topographical surveying is employed, whereas cadastral surveying is adopted for plans specifically indicating boundaries between properties and other associated features. Mining surveying, as the name implies, is purely for mining work.

The surveying for the construction industry is generally referred to as engineering surveying, and hereafter in this chapter, everything mentioned relates to this particular branch. Engineering surveying too has it's sub-divisions, the main ones being triangulation, traversing, tacheometry and levelling, although both photographic and aerial surveying are playing an increasing role nowadays in the preliminary surveys of extensive construction projects.

The two main purposes of surveying in the construction industry are to determine the shape and contours of a tract of land, prior to its development for a project, and to set out buildings or other construction work on the ground as shown on the plan. In the first mentioned process, the details taken in the field are systematically recorded, and later in the office they are plotted to a given scale on drawing paper, in order to produce an accurate plan representing the site.

The boundaries of the site and all the relevant features within them, such as buildings, trees, bushes, marshland, ditches, streams, walling, fences, etc. are clearly shown on the plan, and the undulations of the ground are also depicted, either by contour lines or a level grid. The value of such a plan is that it enables the architect, or civil engineer, to study the overall site conditions and design his scheme to the best advantage. Furthermore, it is possible to calculate the available area quite easily from the plan.

A site survey can be carried out in a number of different ways, depending mainly on the size of the site and the extent to which obstructions may impede the progress of the work. Relatively small sites, having few major obstacles and not warranting a high degree of accuracy, can usually be surveyed satisfactorily with a tape, used in conjunction with other simple equipment. Most tapes consist of a rolled graduated steel band, which can easily be pulled out of a leather or steel case and wound back with a simple winding mechanism when not in use. For less precise measurements a reinforced linen or plastic tape is sometimes employed. Obstructions along a line are usually overcome by simple field geometry. Generally, the area being surveyed is divided into triangles, conveniently arranged so that the boundaries and any important internal fence lines etc. run as close to the side of a triangle as possible. This use of triangles is commonly termed triangulation. The sides of the triangles are set out on the ground, care being taken that they are perfectly straight, and accurately measured with the tape fully extended. The straightness of all lines is achieved by sighting (or boning) from one station to another with ranging rods.

From the principal lines, commonly known as base lines, short right angled offsets are measured with a tape or an offset staff to points on the boundaries or to the other relevant features. In this way, the boundaries etc, can be accurately plotted later. The use of a prism square, a sitesquare, or sometimes a cross staff, ensures that the offsets are truly at right angles to the base line. Further lines, known as tie lines, are frequently taken between known points to give added accuracy to the survey. A compass is sometimes used in conjunction with a triangulation survey, so that the bearing of any important line can be recorded as an additional safeguard.

All measured lengths are carefully entered into the field book, together with as many notes as are considered necessary by the surveyor, to thoroughly establish the position of all points. When measuring on sloping ground, due allowance must be made for the slope, since the plan will interpret all measurements as being horizontal. The use of a clinometer or the method of stepping down and the slope with the tape and plumb line will overcome this problem.

In a survey of this nature, a clear understanding between the surveyor in charge of the operation and his assistants is essential. A strict signalling code should always be adopted, especially where it is difficult, in a high wind for example, for one member of the survey party to hear the other.

As a branch of surveying, traversing is generally regarded quite separately from simple linear surveying, even though many smaller traverse surveys can be adequately carried out with a tape and compass. In most cases, where much accuracy is necessary, this type of surveying is best executed with a theodolite, which is the perfect form of angle measuring instrument. Traversing entails running a series of survey lines close to the perimeter of a forest, lake or the like, this form being known as a closed traverse, or along the course of a river, road or similar continous route, which is termed an open traverse. The measurement of survey lines and their bearings, which are the horizontal angles turned between two known lines of direction, such as North 30° East or South 20° West, is the basic principle of traversing.

Because a closed traverse survey ends where it began – that is at the point known as the origin of the traverse, there are several simple ways of checking the accuracy of such a survey.

One of these checks can be carried out by applying the well known formula, $\Sigma\alpha_i = (n-2) \cdot 2R$, where $\Sigma\alpha_i$ is the total number of degrees of all the internal angles, n is the number of angles and R is 90°.

The plotting of a traverse survey is normally carried out by the co-ordinate method. Each point at the end of a survey line has a latitude and a departure, which is related to the origin of the traverse. The latitude is the distance in a northerly or southerly direction and the departure is likewise in an easterly or westerly direction. A point having a known latitude and departure can be easily plotted.

The operation of measuring distances and heights with a transit theodolite, or even a level, by means of the stadia hairs of the reticule carried by the diaphragm of the telescope is known as tacheometry, or stadia surveying. This is a very expedient method, where absolute accuracy is not essential, but it cannot be recommended for really long survey lines.

In order to measure distances in this manner with an orthodox instrument, it is set up over a point at one extremity of the line being measured, and the levelling staff is held upright on a point at the other extremity. Staff readings are taken where the two stadia hairs cut the staff, and the difference between these, known as the staff intercept, is calculated. To determine the distance between the instrument and staff, where the line of sight is truly horizontal, the staff intercept is multiplied by the stadia coefficient, to which is added the additive constant, where this is necessary.

The formula for the calculation of horizontal distances where the telescope has to be elevated or depressed is slightly modified from that just described. The calculation of the difference in ground elevation between the instrument and the staff position is basically trigonometrical, once the horizontal distance between these points and the angle of inclination or declination from the instrument are established. In some recent instruments it is unnecessary to consider this additive constant at all, and in even more modern tacheometers the calculation of both horizontal distances and vertical heights is made automatically.

The primary function of the commonly employed transit theodolite is the accurate measurement of angles in both the horizontal and vertical plane. For this reason, the instrument is fundamentally made up of a graduated horizontal circle, also known as the lower plate, and a graduated vertical circle. The horizontal circle and its hollow spindle, together termed the limb, are carried by and rotate on the levelling head, which consists of two parallel base plates normally separated by three foot screws. The upper base plate is often known as the tribrach. The lower base plate is holed and threaded for screwing on to the top of a tripod, once it has been firmly planted. Manipulation of the foot screws ensures that the instrument is basically in the level position.

The vertical circle, carrying the telescope with its sensitive bubble tube and the vernier arm, is mounted on the standards, which have a trunnion bearing at the top to accommodate the horizontal, or transverse axis, and a plate level at the bottom for levelling the instrument. The standards are attached at the base to the vernier, or upper horizontal plate. The underside of the vernier plate has a solid inner spindle, which fits into and rotates inside the hollow outer spindle of the limb.

Clamps and tangent screws, also termed slow motions, are provided to fix the position of both the horizontal and vertical circles, as well as their vernier plate and arm respectively. Many vernier pattern transit theodolites are still in regular use, although the principle of the vernier has mainly given way to the optical micrometer. In the same way, the prismatic plumbing device is fast superseding the former plumb bob and shifting plates incorporated in the levelling head.

A theodolite is subject to temporary (or station) and permanent adjustments, the former being a relatively simple procedure carried out by the surveyor in the field. The latter should never be attempted by the surveyor, but left to the manufacturer, so as to avoid irreparable damage to a precision built instrument.

Levelling is simply surveying in the vertical plane to ascertain the varying levels of the ground, primarily so that contour lines or a level grid can be included on the site survey plan. This can be achieved by three differing methods, namely spirit levelling, being by far the most usual in connection with construction sites, angular levelling and hypsometric levelling. Spirit levelling entails the use of a level, of one type or other, and a levelling staff. The operation can normally be carried out quite satisfactorily by the surveyor, using the instrument and booking the readings and notes, and one assistant, named the staffman, holding the staff upright on those points where readings are required.

Before commencing such a levelling operation, a point of known elevation above a fixed datum is normally established as a basis, to which all staff readings can be directly or indirectly related. This point is either a permanent or temporary bench mark, commonly referred to by the abbreviation B. M. or T. B. M. respectively in the level book. A bench mark is a point having an accurately determined elevation, usually as part of a wide spread survey officially carried out by the Ordnance Survey. A temporary bench mark on the other hand, is a point on the top of a concrete road kerb, manhole cover, or the like, the known elevation of which is related to a B. M. In some cases on smaller sites, where a B. M. or T. B. M. is not available within a reasonable distance, an arbitary T. B. M. with a notional elevation of say 100.00 metres is created.

The actual levelling operation commences with a back sight reading from the level on to the staff held on the B. M. or T. B. M. Readings to other points with the level in the same position are termed intermediate sights, and the last sight from that position prior to moving the level, is known as a foresight. From the next position of the level, a backsight is taken on to the staff held in the same position as for the preceding foresight. This common staff position is termed a change point. The line of sight of the instrument set up in any position is known as the collimation height or line, and this alters at each change of position of the instrument. At the end of a levelling operation, or even at the end of each day's levelling, a direct or indirect backsight (via change points) should be taken at the B. M. or T. B. M. as a closing check on the accuracy of the work.

The readings are recorded in the level book in such a way that the calculation of the levels, known as reduced levels, at each point can be rapidly done. There are two simple methods by which the reduced levels are determined, one by the rise and fall and the other by the collimation method, the latter being the more popular with most surveyors. Fig. 17 illustrates a typical double page of a British level book, on which a sketch with notes and dimensions is included as added information.

Several types of level are frequently used on construction sites, three principal ones being the dumpy, tilting, and automatic level. The dumpy level consists basically of a telescope, having a sensitive level tube attached either to the top or side, and a levelling head, similar to that of a theodolite. When correctly set up on the tripod, the line of collimation, or sight, is always at right angles to the vertical axis of the instrument. The tilting level varies from the dumpy, in that the telescope is not fixed at right angles to the vertical axis, but can be slightly elevated or depressed by means of a micrometer screw. Unlike some dumpy levels, this level is always provided with a second spirit level, which is a smaller circular one, known as a pill box bubble or cat's-eye, used to level the instrument approximately. The larger spirit level is checked for accuracy at each individual reading. Sometimes the levelling head is of the ball and socket or concave joint, rather than the screw type, in which case the level is often termed a "quickset". Most forms of tilting level are also known as precision levels.

The automatic (or autoset) level is dissimilar to both the dumpy and tilting level, in that it need only be levelled up with the small circular spirit level. Thereafter, the line of collimation at every reading is automatically levelled by a built-in optical compensator. This instrument is more compact than most other levels and is ideal when accurate readings are required quickly. A FENNEL automatic builders' and

BACK-SIGHT	INTER-MEDIATE	FORE-SIGHT	RISE or COLLIMATION	FALL LINE	REDUCED LEVEL		DISTANCE	REMARKS
Note	Reduced Level of			T.B.M.	on MH			10.00m GRID N
	Cover in	Vine Street		is	26.83		G	TBM x VINE STREET 6.13 WIDE
								HEDGE
							H	FENCE PATH 1.00 WIDE
			COLLIMATION LINE					
							J	1.46 WIDE CLIVE LANE
								17 18 19 20 21 22 23 24
							ALONG GRID	NORTH EAST CORNER OF SITE.
							LINE G	
1.44			28.27		26.83		—	Backsight on T.B.M. (MH Cover in Vine St)
	1.36				26.91		165.20	Point G17 on grass (2.16 South of path edging)
	1.42				26.85		175.20	" G18 on Concrete path edging.
	1.38				26.89		185.20	" G19 on tarmac path.
	1.24				27.03		195.20	" G20 on tarmac road (2.11 North of kerb)
	1.14				27.13		205.20	" G21 " " "
	1.28				26.99		215.20	" G22 " " "
	1.41				26.86		225.20	" G23 on Concrete road kerb.
	1.53				26.76		235.20	" G24 on gravel (3.09 North of kerb)
		2.26	28.27		26.01		—	Change Point between G19 + G20.
0.42			26.43		26.01		—	" " " " " "
	0.71				25.72		LINE H 165.62	Point H17 on grass.
	0.80				25.63		175.62	" H18 " "
								Continued

Fig. 17: Double Page of a British Level Book

engineers' level is detailed in Fig. 18. It provides a bright erect image, is accurate to within about 2 mm over a distance of a kilometer and features a recessed objective lens, which makes a sun shade unnecessary.

Many levels, like theodolites, are provided with a clamp and tangent screw for fixing the position of the telescope. As with a theodolite, a level is subject to temporary and permanent adjustments, and similar rules apply to both instruments.

Probably, above all other surveying instruments, the varying types of conventional theodolite and level have given the most invaluable service to the construction industry for a long time. In their most modern forms – for example, the automatic level, microptic theodolite and self-reducing tacheometer – they have reached a high standard of accuracy and expedience. However, new revolutionary instruments, some still in their infancy, are fast emerging and threaten to replace them on the larger scale surveying operations. Latest developments in this field include electro-optical distance measuring instruments, like "geodimeters" and "tellurometers", and the laser beam instrument.

Office work constitutes a vital part of all surveys, for it is in this department that a good understanding of the field work and correct interpretation of the notes are necessary, if an accurate plan is to be produced. The competent use of the many drawing instruments is also most important. Of all these instruments, those most commonly utilised include compasses (ordinary, spring bow and beam) dividers, tee and set squares, French curves, scales and protractors. Special instruments, such as the pantograph and eidograph for enlarging or reducing plans, and the planimeter for calculating areas on the plan, are also often employed in the surveyor's office.

The lettering and legend on any plan produced on the drawing board should always be clear and concise in giving all the information necessary for a complete understanding of the plan. Well executed field work must be followed up by thorough office work, if it is to be of any benefit to those using the plan.

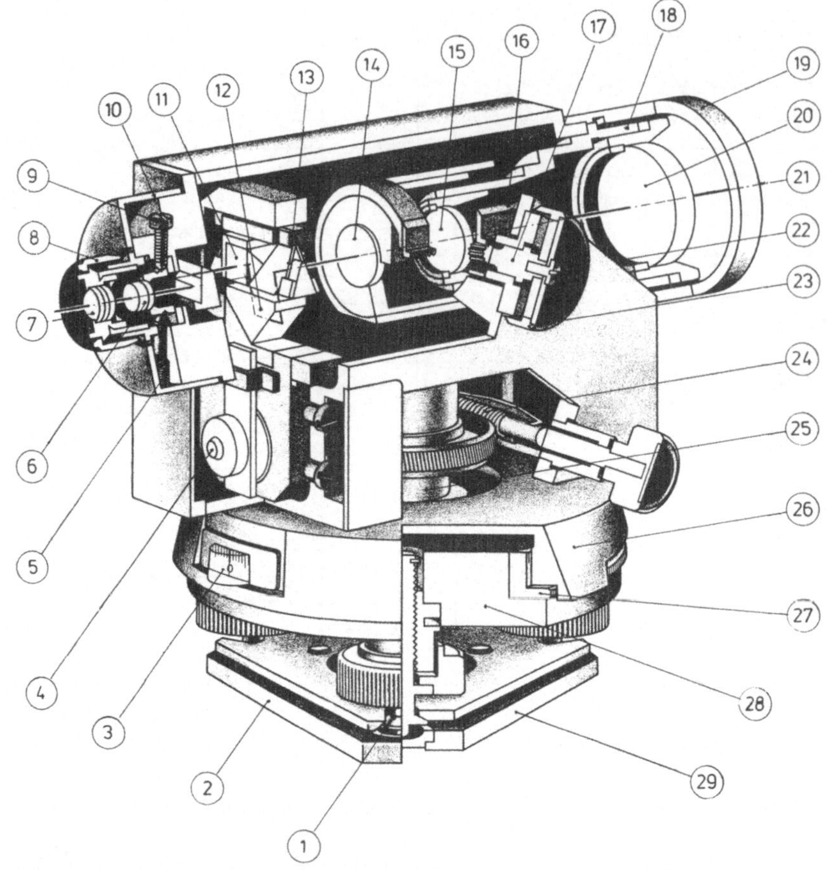

Fig. 18: FENNEL Automatic Builders' and Engineers' Level

1	**levelling screw with knurled perimeter**	Stellschraube mit Rändelknopf	vis de réglage cannelée
2	**base plate**	Grundplatte	planchette de base
3	**index bearing magnifying glass**	Zeigerlupe	verre de lecture grossissant
4	**compensator damping device**	Kompensatordämpfung	ajustage de compensation
5	**coilspring**	Schraubenfeder	ressort hélicoïdal
6	**reticule, reticle**	Strichplatte	plaquette graduée
7	**eye-piece lenses**	Okularlinsen	lentilles de l'oculaire
8	**resetting ring**	Einstellring	molette de mise au point
9	**adjusting screw**	Justierschraube	vis de réglage
10	**eye-piece cover**	Okularkappe	boîtier de l'oculaire
11	**suspended compensator prism**	Kompensator-Pendelprisma	prismes oscillants de compensation
12	**fixed prism**	feststehendes Prisma	prismes fixes
13	**pendulum thread**	Pendelstrang	fil pendulaire
14	**focussing lens**	Einstellinse	lentille de mise au point
15	**negative lens**	konkave Linse	lentille concave
16	**rack**	Zahntriebstange	tringle dentée d'entraînement
17	**pinion shaft with pinion**	Triebwelle mit Ritzel	tringle d'entraînement avec pignon
18	**objective lens tube**	Objektivrohr	tube de l'objectif
19	**objective cover ring**	Objektivschutzrohr	tube protecteur de l'objectif
20	**objective lens**	Objektivlinse	lentille de l'objectif
21	**line of collimation, line of sight**	optische Achse	axe optique

22	**locking ring**	Vorschraubring	bague de blocage
23	**focussing knob**	Scharfeinstellungsknopf	molette de mise au point
24	**horizontal slow motion spindle**	Schraubenfeintrieb für horizontales Drehen	vis de précision pour réglage horizontal
25	**vertical axis**	senkrechte Achse	axe vertical
26	**cover plate**	Abdeckung	protection
27	**horizontal circle**	Teilkreis	cercle horizontal
28	**tribrach (part of the levelling head)**	Dreifuß (Teil des Unterbaus)	partie de la structure inférieure
29	**tension plate**	Spannplatte	planchette de fixation

Vocabulary	Wörterverzeichnis	Glossaire
surveying	Vermessung, Vermessen	levés topographiques
geodetic surveying	Geodäsie, Landvermessung, Feldmessen	**levés géodésiques**
hydrographical surveying	hydrologische (gewässerkundliche) Messungen	levés hydrologiques
topographical surveying	topographische Vermessung (der Gelände- und Bodengestaltung)	levés topographiques
cadastral surveying	Katastervermessung (für das Grundbuch)	mensuration cadastrale
mining surveying	Markscheiderei, Vermessung im Bergbau	levés souterrains
spheroidal shape	kugelige Gestalt	forme sphérique
survey	Geländeaufnahme	levé de terrain
map	Landkarte	carte géographique
to depict, to plot	darstellen, aufzeichnen	**décrire, tracer**
ground feature	Grundrißgestaltung	dessin de base
plan	Plan, Grundriß	plan horizontal
boundary	Grenze	limite
property	Grundstück	propriété
engineering surveying	bautechnische Vermessung	levés pour le génie civil
traversing, traverse survey	Polygonierung	cheminement
triangulation	Triangulieren, Dreiecksaufnahme	triangulation
tacheometry	Tachymetrie	levés tachéométriques
levelling	Nivellieren	nivellement
aerial surveying	Luftvermessung	aéro-photogrammetrie
to set out	abstecken	jalonner
in the field	im Gelände	sur le terrain
to record	protokollieren	inscrire
office	Büro	bureau
scale	Maßstab	échelle
drawing paper	Zeichenpapier	papier à dessin
site	Bauplatz, Baustelle	chantier
marshland	sumpfiges Gelände	terrain marécageux
ditch	Graben	fossé
walling, wall	Mauer	mur
fence	Zaun	clôture
undulation of the ground	Bodenwelle	ondulation de terrain
contour line	Höhenlinie	courbe de niveau
level grid	Höhenrost, Höhenraster(netz)	grille de cotes
to design a scheme	einen Plan entwerfen	dresser un plan

tape	Bandmaß, Maßband	mètre-ruban
graduated	mit Maßskala versehen	gradué
reinforced linen tape	(mit Bronzefäden) verstärktes Textilband	ruban renforcé
by field geometry	m. H. von Hilfsmessungen	avec levés complémentaires
sighting, boning	Visieren, Fluchten	viser
ranging rod	Fluchtstab, Bake	jalon
base line, principal line	Haupt(meß)linie	ligne de base
offset staff	Meßlatte	mire parlante
tie line	Verbindungslinie, Hilfslinie	ligne auxilaire
compass	Kompaß	boussole
bearing	Winkelabweichung (i.a.zur Nord-Süd-Richtung)	déviation
the lengths are entered into the fieldbook	die Längen werden ins Feldbuch eingetragen	les longitudes sont inscrites au registre
sloping ground	abfallendes Gelände	terrain descendant
due allowance	passender Abzug bzw. Zulage	déduction ou adjonction convenable
clinometer	Neigungsmesser	clinomètre
slope	Gefälle	pente
plumb line	Lotschnur	fil à plomb
surveyor in charge	leitender (verantwortlicher) Vermesser	géomètre en chef
signalling code	Signalsystem	code de signalisation
to run a series of survey lines	eine Folge von Meßlinien legen	fixer une série de lignes de référence
closed traverse	geschlossener Polygonzug	polygone fermé
open traverse	offener Polygonzug	polygone ouvert
origin	Festpunkt, Bezugspunkt	point fixe
co-ordinate method	rechtwinkliges Koordinatensystem	système de coordonnées rectangulaires
latitude	y-Koordinate, Ordinate	latitude
departure	x-Koordinate, Abszisse	abscisse
transit theodolite, transit	Theodolit	théodolite
level	Nivellierinstrument	niveau
stadia hairs	Distanzfäden	fils
reticule, reticle	Fadenkreuz, Strichplatte	plaquette graduée
diaphragm	Fadenkreuzträger	porte-réticule
telescope	Fernrohr	télescope
levelling staff	Nivellierlatte	mire
staff reading	Lattenablesung	lecture de mire
line of sight, collimation line	Zielachse, Sehlinie	ligne de collimation
stadia coefficient, stadia constant	Multiplikationskonstante	constante du télémétre
additive constant	Additionskonstante	constante d'addition
inclination	Neigung	inclinaison
declination	Abweichung	déclination
horizontal circle, lower plate	Horizontalkreis	cercle horizontal
limb	Limbus	limbe
base plate	Grundplatte	plateau
foot screw	Fuß(stell)schraube	vis de réglage
tribrach	Unterbau	support
tripod	Stativ	trépied
bubble tube, level tube	Wasserwaage	niveau à bulle d'air
vernier	Nonius	vernier
standards	Fernrohrträger	support de lunette
trunnion	Drehzapfen (der Kippachse)	pivot

bearing	Lager	appui, coussinet
transverse axis, horizontal axis	Querachse, Kippachse	axe transversal
clamp	Feststeller, Arretierung	arrêtoir
tangent screw, slow motion	Feintrieb	dispositif d'entraînement lent
pattern	Modell	modéle
optical micrometer	optisches Mikrometer	micrométre optique
prismatic plumbing device	Schwing-Prismen-Horizontierung	dispositif d'horizon à prismes
plumb bob	(Blei-)Lot	plomb
shifting plate	Verschiebeplatte	plateau de réglage
adjustment	Berichtigung, Einrichten, Justieren	ajuster
site survey plan	Geländeaufnahme(-Plan)	levé de terrain
spirit levelling	Nivellieren	nivellement
angular levelling	trigonometrische Höhenmessung	tachéométrie trigonométrique
hypsometric levelling	barometrische Höhenmessung	nivellement barométrique
surveyor	Vermessungsingenieur	géomètre
reading	Ablesung	lecture
note	(Feldbuch-)Aufzeichnung	note
staffman	Vermessungsgehilfe, Lattenträger	aide-géomètre
fixed datum	Bezugshöhe	cote de référence
basis	Ausgangs(höhen)punkt	base
bench mark (B.M.)	Höhenfestpunkt	repère
temporary bench mark (T.B.M.)	provisorischer Höhenpunkt	repère temporaire
Ordnance Survey	Amtliche Landesvermessung	topographie nationale
road kerb	Bordstein	bordure de trottoir
manhole cover	Einstieg-Schachtdeckel	couvercle de regard
arbitrary T.B.M.	willkürlicher Festpunkt, "Architekten-0-Punkt"	repère arbitraire
notional elevation	Nennhöhe	niveau nominal
back sight(reading)	Rückblick(-Ablesung)	visée inverse
intermediate sight(reading)	Seitenblick-, Zwischenblick (-Ablesung)	visée intermédiaire
foresight(reading)	Vorblick(-Ablesung)	visée directe
change point	Wechselpunkt	point de changement
collimation height, collimation line	Instrumentenhorizont, Höhe der Sehlinie, Zielachse	ligne de collimation
closing check	Schlußkontrolle	vérification
reduced level reduced mean sea level	reduzierte Höhe, Höhe über N.N.	horizon réduit
rise and fall method	Berechnung mit Höhenunterschieden	méthode de dénivellation
collimation method	Berechnung über den Instrumentenhorizont	méthode de collimation
dumpy level	Baunivellier, Nivellierinstrument ohne Kippschraube	niveau à lunette fixe
tilting level	Nivellierinstrument mit Kippschraube	niveau à lunette avec vis de basculement
automatic level, autoset level	Nivellierinstrument mit automatischer Horizontierung, Nivellierautomat, Kompensationsnivellier	niveau automatique
micrometer screw	Mikrometerschraube	vis micrométrique
pill box bubble, circular spirit level, cat's eye	Dosenlibelle	niveau sphérique
ball and socket joint, concave joint	Kugelgelenk, Gelenkkopf	articulation à bille

precision level	Feinnivellierinstrument	niveau de précision
microptic theodolite	Theodolit mit Skalenmikroskop	théodolite microptique
self-reducing tacheometer	Reduktionstachymeter	tachéomètre autoréducteur
field work	Außentätigkeit	travaux extérieurs
drawing instrument	Zeichengerät	outil à dessiner
compass	Zirkel	compas
spring bow compass	Teilzirkel	compas d'épaisseur
beam compass	Stangenzirkel	compas à calibrer
dividers	Stechzirkel	compas à pointes sèches
tee square	Reißschiene	té
set square	Zeichendreieck	équerre
French curve	Kurvenlineal	pistolet
scale	Maßstab	échelle
protractor	Winkelmesser, Transporteur	rapporteur d'angles
lettering	Beschriften, Beschriftung	inscription
legend	Zeichenerklärung	légende
drawing board	Zeichenbrett	planche à dessin

Dialogue

Robert: In Britain, the so-called surveying profession seems to cover a wide field, and therefore, it is sometimes rather confusing for a Continental, like myself, to appreciate what a surveyor's responsibilities really are. For example, during my many visits to England I've often come across such titles as "Borough Surveyor" and "Valuing Surveyor". Neither of these, Duncan, I imagine, have anything to do with the construction industry.

Duncan: The two types of surveyor that you mentioned are not often directly associated with new construction projects. A Borough Surveyor is a public servant employed by a local authority to supervise public services, amongst other duties, of a particular district. On the other hand, a valuing surveyor is principally involved in assessing the value of existing properties for buying and selling. I think we would be drifting away from our real subject, Robert, if we continued on these lines, so perhaps we should only briefly discuss the functions of those surveyors who are directly concerned with construction.

Robert: Yes, I'm sure you are right, and maybe the best way would be to begin by telling our readers the basic differences between a quantity surveyor and a builder's surveyor. I was surprised to learn, I must confess, that there are subtle differences.

Duncan: Well, the fundamental difference between the two is that the quantity surveyor, commonly referred to as the Q.S., is appointed to prepare bills of quantities and final accounts etc. and generally look after the client's interests. The builder's surveyor, on the other hand, is employed by the contractor to interpret the bills, keep an eye on the cash flow, and protect his interests. I must also add here that both should have a fair knowledge of land surveying, because it plays some part in their work. With the ever increasing and more professional responsibilities of a contractor's surveyor these days, Q.S.s, rather than just builder's surveyors, are generally being employed by British contractors.

Robert: Putting two and two together, it's obvious that the two must often work together and come to some sort of agreement on problems in the measurement of construction work and in the financial implications of a contract.

Duncan: Broadly speaking, this is so, and having worked as both a Q.S. and a builder's surveyor, I can assure you that a good sense of compromise is always necessary.

Robert: I know if I give you too much rope on this particular topic, of which you are doubtless well informed, you'll ramble on and on, so shall we move on to something else?

Duncan: I take your point, Robert, so maybe I can clarify any other points concerning surveying in Britain, which you feel would interest our readers.

Robert: Yes indeed, I do have a point about surveying instruments. Looking at a recent British catalogue, I was amazed to find reference to such old items as chains and plane tables amongst other far more up-to-date equipment. Do you really use these?

Duncan: I know you reckon that chains went out with "The Arc" and I wouldn't argue that they are somewhat antiquated. In fact, very few are in use in Britain today, although they are still available. In my thirty odd years in the profession, to be perfectly honest, I've only handled a chain on a few occasions, and even then it was normally a steel band chain, rather than the older chain made up of steel wire links. Of course, steel or invar (that is nickel steel alloy) tapes have largely replaced the rather clumsy chains.

Robert: I assume that chainmen were surveyors who worked with chains, or am I on the wrong track?

Duncan: Well it's a reasonable assumption, but not strictly correct. They handled chains and other primitive equipment, like ranging rods and offset staffs, in order to take linear measurments, but were normally assistant or junior surveyors.

Robert: You certainly won't see any chains here on the Continent, and we aren't all that familiar with the plane table either.

Duncan: Here again, the plane table is seldom used in Britain nowadays, and in any case, it's only really practical where the weather is reliable. Incidentally, even though the plane table is no new idea, there are many modern versions now available.

Robert: Can you please give me some inkling, Duncan, of the general principle of the plane table?

Duncan: Well, it's a very simple idea, whereby a drawing board on a tripod is combined with an alidade, which is basically a telescope and often the same as the upper part of a transit theodolite.

Robert: As I understand it, the big advantage of the plane table is that a drawing representing the site can be produced on the spot. In good weather conditions, I can imagine this system has wide possibilities and is a very expedient form of surveying.

Duncan: Yes that's quite true. Perhaps I could go a step further and broadly explain the use of the plane table.

Robert: Please do, because I'm sure it could be of some interest, especially to anyone likely to work in hot countries, like South Africa or India, for example.

Duncan: Once the plane table board is accurately set up over a station in the field and the alidade orientated in the direction of another station or other important feature, a line is simply drawn along the edge of the alidade rule, thus fixing the direction on the plan. This procedure is repeated with the plane table over other positions, and in this way all the stations and distances are plotted, and thus a complete plan is drawn to scale.

Robert: Thanks for putting me in the picture. As you know, greater use is being made these days of upright image instruments, which I feel have great advantages. What are your thoughts?

Duncan: I must agree this is so, but what with this gradual change from inverted image instruments plus the changeover in Britain from the imperial to the metric system of measurement, things were a little hectic.

Robert: I trust you're not complaining about these logical changes.

Duncan: Not really, but for a while we did experience an awkward period in Britain during the wind of change. It might be surprising to you, but often I had to switch from one system of measuring to the other in the same day, besides using instruments of varying pattern.

Robert: Actually, Duncan, I can appreciate that you had problems. I know that the imperial system has now died a natural death, but with so many inverted image instruments still being produced, because of their lower cost, both you in Britain and ourselves will have to live with a variety of instruments for a while yet.

Duncan: On the other hand, Robert, more expensive instruments, like the automatic level, for example, are fast coming into use. Incidentally, which form of compensator in this type of level do you prefer?

Robert: To be candid, I find it difficult to say. The two types with which I am most familiar, that is the suspended prism or mirror type and the moving bubble type, both have their advantages. By the way, within what limits of inclination and depression will the compensator function effectively?

Duncan: I would say about 15 minutes of arc above and below the horizontal, which means that the surveyor has only to set up the instrument very roughly and quickly.

Robert: I am rather interested to know if any instruments manufactured in Britain are provided with 400^g rather than a $360°$ circles.

Duncan: As yet I haven't seen very many but I'm pretty sure that Continental products generally are moving far more rapidly towards the circle of 400 new °, known as 400^g, than we are in Britain. As in so many other fields, I'm afraid we tend to be a little stick in the mud about change.

Robert: Moving even more up-to-date, do you consider that the modern electronic measuring instruments, like those working on the laser beam principle for instance, have any real future on surveys for construction work?

Duncan: Not on the small construction site at the moment, because the high initial cost and expense of maintaining such intricate equipment are rarely justified, but they certainly have prospects for larger sites. It must be said, however, that they are very precise in measuring over long distances.

Robert: You are absolutely right, and this is of particular value, when using a geodimeter for example, in the inspection of bridges, harbours and dams, to determine small displacements.

Duncan: It is also worth mentioning that the high intensity beam of such an instrument makes its use possible in rainy and misty weather. This must be one of its principal advantages over the theodolite.

Robert: On the other hand, one major disadvantage, in my opinion, is that a separate power supply is necessary, which makes the equipment rather heavy and cumbersome.

Duncan: I agree, but sooner or later manufacturers will surely design far more compact and much lighter instruments, just as they have in the past with similar products, such as photographic gear.

Robert: You're probably right, and as a result, their cost could become less prohibitive. As you know, we already employ many outstanding automatic optical instruments on Continental sites, and these have now developed into compactly and solidly built equipment, being easy to handle.

Duncan: I believe that any instrument which provides improved accuracy and easy handling to speed up the surveying work is well worth investing in, even if it is somewhat more expensive.

Robert: As another discussion point, maybe we could enlarge a little on photographic and aerial surveying, which we only touched on in our chapter. Photography, of course, is common to both, even though each branch is regarded as being distinct from the other. Duncan, where do you think each one is most suitably employed?

Duncan: Well, I understand that the most satisfactory results from photographic surveying are usually obtained on deep rugged terrain, free from forest or bushland, but having steep slopes.

Robert: The type of country, in fact, which would make aerial surveying unsatisfactory.

Duncan: Exactly. In such a situation, clear photographs from various angles carefully recorded can be of some real value when the survey is being plotted.

Robert: Now, what about aerial surveying?

Duncan: I was coming to that, Robert. This has its main advantages in thickly wooded country or swamp land of constant elevation, where normal land surveying would be difficult, if not nigh impossible, and ground photography would give inadequate results.

Robert: I can appreciate, Duncan, that you're not entirely at home with this topic and I won't press you for more explanations. However, getting back to everyday surveying, what do you think about the split-bubble system adopted for modern levels.

Duncan: I think it's ingenious. By the way, it's also known as the coincidence system.

Robert: Perhaps you could say a word or two about it then?

Duncan: Yes, of course. The idea employs a system of prisms, which appears to split the bubble in half lengthways and reflects the two split ends into the viewing microscope, located near the telescope eyepiece. When the two split ends appear side by side to form a smooth curve, which is termed the coincidence setting, the bubble is centred.

Robert: What advantage has this system over the open tubular scale system?

Duncan: Well, it's common knowledge that the accuracy of any spirit level depends on the sensitivity of the level bubble, as well as on the magnification of the telescope, and the centring of a sensitive bubble is more accurately carried out with the split-bubble system.

Robert: You mentioned magnification, which not only determines the accuracy of staff readings and estimations, but also controls the maximum sighting distance. It should be remembered, however, that a telescope of high magnification must have a larger clear objective aperture than that of a normal telescope, in order to give a wide enough field of view to ensure a satisfactory image brightness.

Duncan: Yes, that's an important point at which we can aptly bring this discussion to an end.

Vocabulary	Wörterverzeichnis	Glossaire
Borough Surveyor	Baupolizeibeamter (Bauüberwachung)	inspecteur municipal des travaux
Valuing Surveyor	Schätzer, Gutachter, Sachverständiger	expert taxateur
public servant	Beamter	fonctionnaire
local authority	Behörde, Amtsstelle	administration publique locale
public services	öffentliche Versorgungseinrichtungen	services publics
public service	Staatsdienst	service public d'état
quantity surveyor, (Q.S.)	Ingenieur, der Bauten finanziell abwickelt, vom Bauherrn beauftragt	calculateur
builder's surveyor, contractor's surveyor	dto., jedoch vom Bauunternehmer beauftragt	calculateur en bâtiment
bills of quantities	Leistungsverzeichnis	tableaux des quantités de matériaux
final account	Schlußrechnung	décompte final
client	Bauherr	**propriéteur**
contractor	Bauunternehmer	entrepreneur
cash flow	Überschuß an Flüssigkeitsmitteln	cash flow
land surveying	Feldmessen	géodésie
putting two and two together, (to draw one's own conclusions)	seine Schlüsse ziehen	tirer ses conclusion
to come to an agreement	sich verständigen	parvenir à un accord
implication	tieferer Sinn	suite impliquée
contract	Vertrag	contrat
to give someone too much rope, (to give someone too much freedom)	jemandem freien Lauf lassen	laisser trop de liberté à quelqu'un
to ramble on and on (to keep on talking)	sich (uferlos) in Reden verlieren	parler beaucoup pour ne rien dire
to take someone's point, (to understand what someone means)	jemandes Meinung billigen	comprendre et partager l'opinion de quelqu'un
to have a point, (to have something to mention)	etwas vorzubringen haben	avoir quelque chose à exposer
chain	Meßkette	chaine
plane table	Meßtisch	planchette de mesure
went out with the Arc, (to be extinct or very) old fashioned)	gehören der Vergangenheit an	appartient au passé
thirty odd years	etwas mehr als 30 Jahre	un peu plus de trente ans
steel band chain, steel band	Stahlmeßband	chaîne d'arpenteur
steel wire link	Kettenglied aus Stahldraht	maille de chaîne
alloy	Legierung	alliage
junior surveyor, (young surveyor learning the profession – after about 22 years of age he becomes an assistant surveyor)	Vermesser, noch in der Ausbildung	géomètre apprenti

alidade	Alhidade, Zeigerkreis	alidade
site	Bauplatz, Baustelle	site
upright image	aufrechtes Bild	vue verticale
inverted image	auf dem Kopf stehendes Bild	vue inversée
suspended prism	Schwingprisma	prisme oscillant
moving bubble type	Blasenkompensator-Typ	type à bulle mobile
limit	Grenze	limite
minute of arc	Bogenminute	minute d'arc
initial cost	Anschaffungskosten, Kaufpreis	coût d'achat
displacement	Lageveränderung, Verschiebung, Setzung	déplacement
high intensive beam	hochwirksamer Strahl	rayon à haute intensité
power supply	Energieversorgung	alimentation en force
photographic gear	photographisches Gerät	appareil photographique
swamp land	sumpfiges Gelände	terrain marécageux
split-bubble system, coincidence system	Koinzidenzlibellen-Anordnung	système à bulles coïncidantes
viewing microscope	Ableseokular	microscope à lecture directe
telescope eyepiece	Okular des Fernrohrs	oculaire de télescope
setting	Einstellung	réglage
open tubular scale	freiliegende (Libellen-) Rohrteilung	graduation tubulaire libre
sensitivity	Empfindlichkeit	sensibilité
magnification	Vergrößerung	agrandissement
staff readings and estimations	Lattenablesungs- und Schätzwerte	lecture de latte et estimation
maximum sighting distance	größte Zielweite	distance maximale de visée
objective aperture	Objektöffnung	ouverture d'objectif
field of view	Sehfeld	champ de vue
image brightness	Bildhelligkeit	clarté d'image

9. PLANNING AND COST CONTROL OF A CONTRACT

The planning or programming of the work is the first essential step in the execution of a contract, and in many cases the contractor is required under the contract conditions to produce a programme. The basic principles of such a programme must be to the satisfaction of the architect, or civil engineer, and should also be approved by the consultants dealing with specialised work, such as heating, electrical and lift installation. It is fundamentally in the contractor's own interest to plan his contract, besides the fact that a detailed programme of work over the contract period does indicate to all concerned the anticipated position of the contract at any given time. Should the contract be running behind schedule, for example, this situation will soon become apparent by referring to the programme.

The resources which a contractor has available – that is his capital, manpower, plant, administration etc. – largely determine his programme of work, and it is important that these should be neither underemployed nor overstretched. In order to use the resources to the best advantage, most large and some medium sized contractors employ a planner, who is also known as a planning officer in some companies. His main responsibility is to plan the execution of the contract within the stipulated contract period, draw up a programme and ensure that the programme is properly put into operation. Consultations between the planner, contract manager and foreman, all of whom will be concerned in the running of the contract, are necessary to produce a workable programme. There are various forms of programme, and two commonly adopted ones are the bar chart and network. Fig. 19 and 20 illustrate parts of a typical form of British bar chart and network respectively.

CONTRACT: HYDE VALLEY WATER TREATMENT WORKS — JOB. No. 2621.

ARCHITECT:	CONSULTANTS:	QUANTITY SURVEYOR:	CONTRACTOR:
JONES & TAIT A.R.I.B.A.	MECHANICAL – HALL & BYE A.M.I. Mech. E. ELECTRICAL – MULLET & PTNS. A.M.I.E.E.	JOHN S. RICE. F.C.S.I.	M.W. McBRIDE LTD.

No	ELEMENT	WEEK COMMENCING (1972)	8/5	15/5	22/5	29/5	5/6	12/6	19/6	26/6	3/7	10/7	17/7	24/7	31/7
		WEEK No	1.	2.	3.	4.	5.	6.	7.	8.	9.	10.	11.	12.	13.
1.	GENERAL EXCAVATION (INCL BASEMENT)	ANTICIPATED DURATION													
		ACTUAL DURATION													
2.	HARDCORE FILLING AND BEDS	ANT. DUR.													
		ACT DUR													
3.	CONCRETE FDNS AND BASEMENT SLAB	ANT DUR													
		ACT DUR													
4.	CONCRETE WALLS & PUMP BASES IN BASEMENT	ANT DUR													
		ACT DUR													
5.	MECHANICAL INSTALL. TO PUMPS ETC	ANT DUR													
		ACT DUR													
6.	DRAINAGE & B/W TO MECHANICAL & ELECTRICAL INSTALL.	ANT DUR													
		ACT DUR													
7.	CONCRETE SUSP SLAB OVER BASEMENT & EXT CONCRETE TANKS	ANT DUR													
		ACT DUR													
8.	BRICKWORK, STONEWORK & BLOCKWORK GENERALLY	ANT DUR													
		ACT DUR													
9.	CARPENTRY 1ST FIX	ANT DUR													
		ACT DUR													

Fig. 19: Part of one Typical Form of British Bar Chart

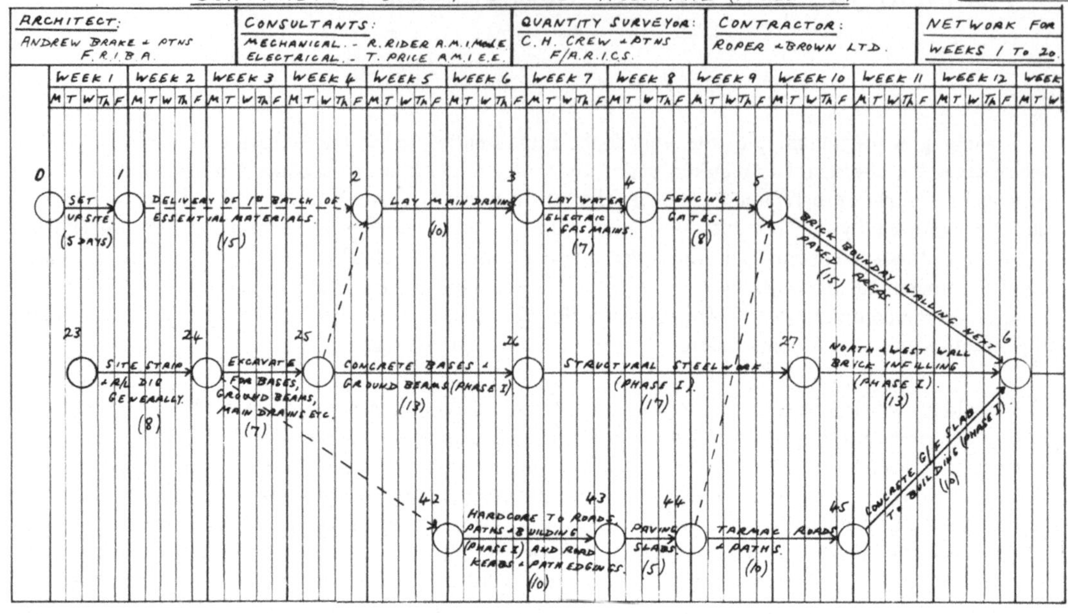

Fig. 20: Part of one Typical Form of British Network

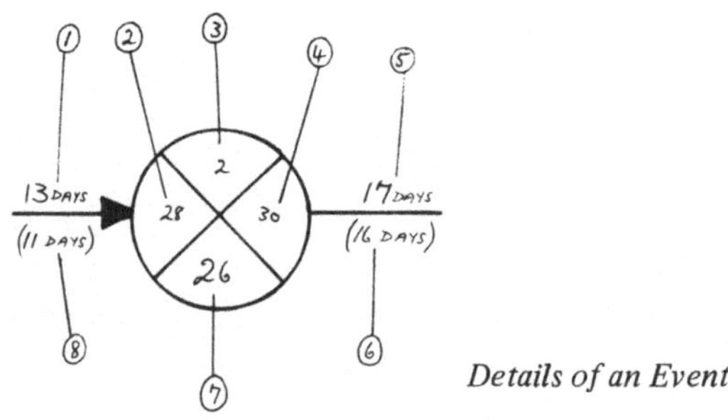

Details of an Event

1	**latest duration of activity (25–26)**	Späteste Dauer des Teilprozesses (25–26)	dernière durée du processus partiel (25 à 26)
2	**earliest event time (28th day)**	Früheste Ereigniszeit (28. Tag)	instant le plus tôt de l'événement (28ème jour)
3	**slack or float of event 26. (2 days)**	Ereignisspiel 26 (2 Tage)	jeu de l'événement 26 dans le temps (2 jours)
4	**latest event time. (30thday)**	Späteste Ereigniszeit (30. Tag)	instant le plus tardif de l'événement (30ème jour)
5	**latest duration of activity (26–27)**	Späteste Dauer des Teilprozesses (26–27)	durée la plus tardive de l'événement (26 à 27)
6	**earliest duration of activity (26–27)**	Früheste Dauer des Teilprozesses (26–27)	durée la plus précoce de l'événement partiel (26 à 27)
7	**event number**	Ereignisnummer	numéro de l'événement
8	**earliest duration of activity (25–26)**	Früheste Dauer des Teilprozesses (25–26)	durée la plus précoce de l'événement partiel (25 à 26)

Abbreviations in Figs 19 and 20

F.R.I.B.A.	Fellow of the Royal Institute of British Architects.
A.R.I.B.A.	Associate Member of the Institute of British Architects.
A.M.I.Mech.E.	Associate Member of the Institute of Mechanical Engineers.
A.M.I.E.E.	Associate Member of the Institute of Electrical Engineers
F/A.R.I.C.S.	Fellow and Associate of the Royal Institution of Chartered Surveyors.
F.C.S.I.	Fellow of the Construction Surveyors' Institute

LTD	limited	B/W	builder's work	G/F	ground floor
PTNS	partners	EXT	external	M	Monday
INCL	including	R/L	reduced level	T	Tuesday
FDNS	foundations	R/L DIG	reduced level digging	W	Wednesday
INSTALL	installation	SUSP	suspended	Th	Thursday
TARMAC	tarmacadam	1ST	first	F	Friday

The bar chart is a simple and easily followed programme, which normally uses bars to represent each element of the contract, like concrete strip foundations, carpentry first fix, glazed screens, or suspended ceilings for instance. The length of a bar indicates the anticipated duration of each particular element. This form of programme is usually adequate enough for the smaller uncomplicated contracts, and can often prove useful as a preliminary or stage plan on larger contracts. Its greatest weakness is that it shows no inter-relationship between one element and another, and thus it is not possible to see which processes of work are totally dependent on others.

A network plan, which is really a logical schematic picture of a project, is far more comprehensive and requires a greater understanding than a bar chart. Its main purpose is to control as well as to plan a contract. It provides the answers to a number of important questions, such as how long the entire work will take, when each job will start and finish and how long any job can be safely delayed without affecting the completion date. It is normal practice to prepare a network in draft initially, so that any subsequent adjustments can be made where necessary. The knowledge of the planner, contract manager and foreman, amongst others, are usually pooled together before any final decisions are made regarding the most economical use of the available resources and the realistic time required to execute each individual job.

The normal form of network is basically made up of a series of events, known also as nodes or stages, which are commonly depicted by circles or other symbols and connected by arrowed lines to show the various sequences of events. These lines, when drawn unbroken, indicate the activities or jobs, such as excavation of stanchion bases, concrete bases to heating ducts, dwarf boundary walling, fencing and gates etc. Events from which activities commence are termed tail events and those at which activities end are termed head events. Broken, or dotted arrowed lines are known as dummies, which rarely have a time value and do not represent actual work. Their main purpose is to link unconnected events logically, so that one cannot possibly occur until the other has been completed. They are also used to a lesser extent to indicate where certain resources are limited, thus affecting the order of events.

Once a network has been drawn up, it is thoroughly analysed to provide much useful information. The initial analysing process is done in two main stages. Firstly, both the earliest and latest dates on which every event can occur is calculated, and from this the spare time on activities and at events can be ascertained. The difference between the earliest and latest date of an event is known as the slack or float of that event and is, in fact, the amount of time the event could be delayed without disrupting the programme. The critical path of any network is the path passing through events having no slack, and the activities forming this path are known as critical activities. The slack path, on the other hand, is that passing through events having slack, and the activities on this path are known as slack activities.

The second stage is the preparation of schedules, which are lists of activities showing their anticipated and critical starting and finishing dates, and thus their expected and permissable duration. In other words, it is the process of assigning calendar times to all the activities. The calculation of the float, on those activities where it occurs, is also given on the schedules. The portion of the float which can be expended without reducing the duration of any succeeding activities, where this is possible, is termed free float, and where the float can be entirely expended without affecting the earlier or later activities, it is termed independent float. Details of both free and independent float are usually given on the schedules in addition to total float.

Setting up the site is an essential and time consuming part of a contract, which should not be overlooked in the overall plan. This entails the siting of such temporary buildings as offices, which are for the use of the contractor's site staff and the architect, civil engineer or clerk of works, messrooms and toilets for the operatives and storage sheds for certain materials. Temporary fencing and hoardings are also often required to prevent trespass on the site and to enclose a compound for materials and items of plant. The dismantling and clearing away of temporary structures etc, and the cleaning up of the site on the completion of the contract are other time consuming items, which must be allowed for in the programme, even though they may seem relatively unimportant.

Putting a programme into actual operation requires a great deal of co-ordination of the contractor's resources, including indirect ones like nominated and domestic sub-contractors and suppliers of materials. Nominated sub-contractors and suppliers are those specifically selected by the architect or civil engineer for specialised sections of the contract. Domestic sub-contractors and suppliers, on the other hand, are those selected by the contractor for normal work. A certain degree of flexibility in programming is necessary to allow for indirect resources, which are often indeterminable at the planning stage.

During the course of most contracts the contractor has to revise his programme from time to time. This process is termed updating, by which the programme is adjusted to comply with changed circumstances. A permitted extension of the contract period, normally due to additional work being ordered, is one reason for updating a programme. Excessive inclement weather or other occurances beyond the control of the contractor are further justifications for an extended contract period. Updating is sometimes necessary as a result of the contractor realising in practice that his original programme is unworkable, due to some miscalculation on his part. Any unjustifiable delay through the contractor's own failure will also necessitate the updating of his programme. This will mean shortening all or some of the remaining activity durations, usually by stepping up the labour force, to make up the time lost. A properly organised contractor, having adequate and efficient resources, is often able to programme his contract to include a possible delay contingency – that is, provided the client allows him a reasonable contract period.

Just as the planning of a contract is important, so too is its cost control. Without such control, management is unable to accurately determine whether the contract is running at a profit or loss. Furthermore, taking the necessary steps at the correct time to curb any loss, or to improve the profit margin, would be virtually impossible. The methods adopted for cost control vary from one contract to another, but the aim in each case is fundamentally the same, that is to know at any time what the financial state of a contract is.
Cost control can be effectively linked to a network programme, since each activity, being a convenient unit of work, can be easily budgeted. Where the rates in the priced bills of quantities are broken down into their various elements – that is labour, material, plant etc. – an estimated cost for each element can be included in the budget figure for the activity, and this is of particular value in the control of costs. This process by the estimator of breaking down each rate is known a analytical estimating. The actual cost of each activity when it is completed on site will clearly indicate how accurate the estimated budget figures are and whether the remaining activities require updated budgeting.

Whatever system of cost control a contractor employs, it is always very closely associated with his cash flow. This is a general term usually referring to the money received through progress payments following interim valuations. Generally speaking, a healthy position is achieved when the cash received for a period exceeds the total cost of wages, payments to sub-contractors, materials, plant hire charges etc., plus on-costs and overheads for the same period by an adequate margin – in other words, when a reasonable profit is being made.

A close check on the financial situation is necessary at regular intervals, and this is usually carried out by the contractor in the form of a cost comparison, which is normally prepared in Britain by the contractor's surveyor as part of his duties. It is simply a comparison between income and expenditure up to a certain point in time, and it is important to ensure that the two are properly related to the same portion of the work executed. Such a comparison can reveal many common contractual failings, like excess wastage of manpower, plant and materials, for example, but if prompt action is taken, many of these can be minimised before they prove too costly.

In keeping with modern trends, much of the work of analysing a large network of say 150 or more activities to provide all the necessary information is done by electronic computers nowadays. This is far less tedious than manual computation, particulary where frequent updating is necessary. The input data is usually prepared by the planner on sheets, transferred to punch cards and fed into the computer. The output data can often be produced in various printed forms to suit individual requirements. Cost control too on large projects is mainly handled by computers. In such work the quick speed sorting and large memory of the machine can be used to great advantage.

Keeping accurate site records, such as those given on the foreman's daily diary and operation sheets can be of immeasurable value for the planning of future contracts. The information included there shows, amongst other things, how long each operation really takes on site and the actual amount of labour required. The feeding of this practical information from the site to the office staff is known as feedback. With regard to costs, feedback is particularly useful to an estimator as a basis for pricing the items in future bills of quantities.

Fig. 21 shows, in the form of a family tree, most of the people concerned in the running of a British building contract employing about 60 operatives. It gives some indictation of how the various responsibilities are distributed and linked to each other. Even though the contractor actually plans the contract, the client's professional representatives often have some influence on the success of the plan, and therefore close co-operation between them and the contractor is essential.

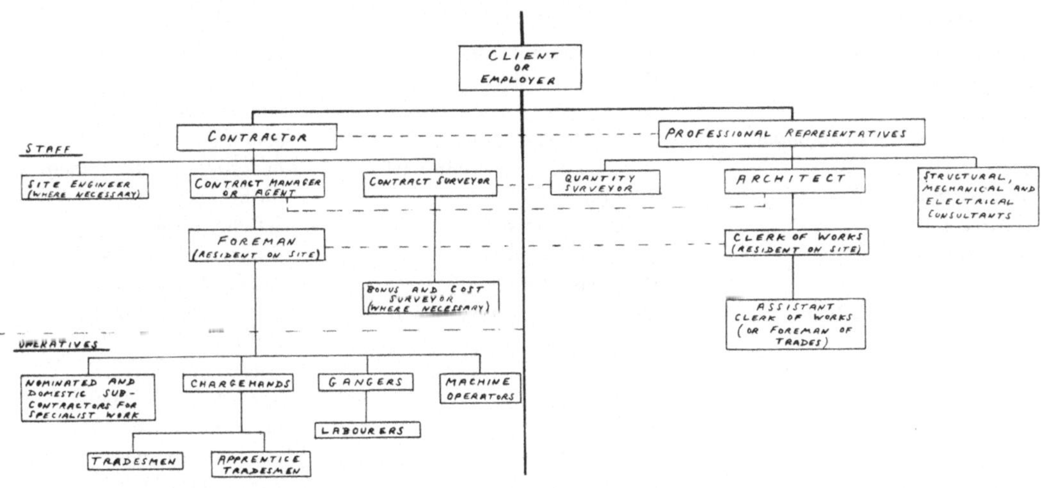

Fig. 21: Probable Site Family Tree for a British Building Contract

Vocabulary	Wörterverzeichnis	Glossaire
planning (or programming) of work	Arbeitsvorbereitung	préparation du travail
cost control	Kostenüberwachung	contrôle des frais
contract conditions	Vertragsbedingungen	conditions du contract
civil engineer	Ingenieur des Ingenieurbaus, „Tiefbauingenieur"	ingénieur du génie civil
consultant, consulting engineer	beratender Ingenieur, (Sonderfachmann)	ingénieur conseil

heating installation	Heizungsbau	installation de chauffage
electrical installation	Elektroinstallation	installation électrique
lift installation	Aufzugsbau	installation d'ascenseurs
contract period	Vertragsdauer, Auftragsdauer	durée du contract
to run behind	zurückbleiben	être en retard
resources	Mittel, Hilfsmittel	ressources
schedule	Bauprogramm, Bauzeitenplan	plan d'avancement
manpower, labour force	Arbeitskraft	main d'œuvre
plant	Baumaschinen und -Geräte	engins et machines
administration	Verwaltung, Hauptbüro	administration
overstretched	überfordert	surmené
planner, planning officer	Arbeitsvorbereiter	chargé de la préparation du travai
contract manager, agent	Bauführer (der Baufirma)	chef de chantier
foreman	Polier	contremaître
bar chart	Balkenplan	plan d'étrésillonnement
network	Netzwerk	treillis
element, section	Teilleistung, Arbeitsabschnitt	élément, section
concrete strip foundation	Betonstreifenfundament, Bankett	fondation continue en béton
carpentry first fix	Zimmerrohbauarbeiten sowie Tür- und Fensterrahmen	gros-œuvre de charpente et menuiserie
(carpentry second fix)	(Türblätter, Fensterflügel Holzverkleidung, Scheuerleisten u. ä.)	(second œuvre de menuiserie)
glazed screen	gläserne Zwischenwand	cloison vitrée
suspended ceiling	abgehängte Decke	plafond suspendu
preliminary plan (stage plan)	Startprogramm (Kurzzeitplan)	plan initial
completion date	Fertigstellungstermin	date d'achèvement
in draft	im Konzept	dans le projet
to pool together, to combine	in Einklang bringen	combiner
event, node, stage	Vorgang, Knoten, Teilprozeß	opération, nœud, tranche
arrowed line	Pfeillinie	flèche
activity, job	Tätigkeit, Aktivität	activité, emploi
stanchion base	Stützenfundament	fondation de poteau
heating duct	Heizungskanal	canalisation de chauffage
dwarf boundary walling	niedere Umgrenzungsmauer	murette de clôture
fencing	Einzäunung	clôture
gate	Tor	porte
tail event	Endereignis	opération terminale
head event	Anfangsereignis	opération initiale
broken line	gestrichelte Linie	ligne interrompue
dotted line	punktierte Linie	ligne pointillée
dummy	Scheintätigkeit, Nullverbinder	faux, attrape
slack, float	Schlupf, Spiel, Puffer	jeu, flottement
critical path	kritischer Weg	chemin critique
to assign	festsetzen, festlegen	déterminer
free float	freie Puffer	flottement libre
independent float	selbständige Puffer	flottement indépendant
total float	totale Puffer, Gesamtspiel	jeu total
to set up the site	die Baustelle einrichten	installation du chantier
overall plan	Gesamtplanung	plan d'ensemble
siting	Errichten, Aufstellen	installation
office	Büro	bureau
contractor's site staff	Baustellenangestellte des Unternehmers	employés cadres de l'entrepreneu
clerk of works	Bauleiter des Bauherrn	chef de chantier
messroom	Speiseraum	réfectoire

operative	Arbeiter	ouvrier
storage shed	Magazin	magasin
hoarding	Bauzaun, Bretterzaun	clôture en planches
trespass	unbefugtes Betreten	entrée interdite
compound	umzäunter Bereich	terrain clôturé
item of plant	einzelne Baumaschinen und -Geräte	divers engins et machines
to dismantle	niederreißen, abbrechen	démolir
to clear away	abräumen	déblayer
temporary structure	Hilfsbauten	constructions provisoires
to clean up	aufräumen, in Ordnung bringen	nettoyer
sub-contractor	Subunternehmer, Nachunternehmer	sous-traitant
supplier	Lieferant	fournisseur
indirect resources	Hilfsarbeiten, Nebenarbeiten	travaux auxiliaires
to update	dem gegenwärtigen Stand anpassen, modernisieren	moderniser
to comply with	sich halten an	se conformer à
extension	Erweiterung	agrandissement
to step up, to increase	vermehren	renforcer
labour force	Arbeitskraft	main d'oeuvre
to make up	wieder gut machen, einholen	rattraper
delay	Verzug, Verzögerung	retard
contingency	Möglichkeit, Reserve	éventualité, possibilité
delay contingency	Zeitreserve für Verzug	réserve de temps pour retard
management	Geschäftsleitung	management
to curb the loss	dem Verlust Einhalt gebieten	arrêter les pertes
profit margin	Verdienstspanne	marge bénéficiaire
to budget, to cost,	veranschlagen,	établir un budget
to make financial allowances	finanziell einplanen	implantation de fonds
rate	Preis	prix
bills of quantities	Leistungsverzeichnis	tableau des quantités, liste de quantités
priced bills of quantities	L. V. mit eingesetzten Preisen, Kostenanschlag	tableau des quantités de matériaux avec prix
labour cost	Lohnkosten	frais de main d'oeuvre
material cost	Stoffkosten	frais de matériaux
plant cost	Maschinen- und Gerätekosten	frais de matériels
estimated cost	kalkulierter Preis	prix calculé
budget figure	Kostenvoranschlagsumme	montant du devis
estimator	Kalkulator	calculateur
to break down a rate	einen Preis aufschlüsseln	détailler un prix
analytical estimating	Preisermittlung aus Einzelfaktoren	prix calculé par analyse
updated budgeting	auf den neuesten Stand gebrachte Veranschlagung	prix mis à jour
cash flow	Überschuß an Flüssigmitteln	cash flow
progress payment	Abschlagszahlung	paiement partiel en fonction de l'avancement
interim valuation	Kostenermittlung für Abschlagszahlung	estimation provisoire
wage	Lohn	salaire
plant hire charge	(Fremd-)Gerätemiete	frais de matériels loués
on-costs	(gesetzliche und tarifliche) Sozialaufwendungen	frais sociaux
overheads	„Feste Kosten", (Allgemeine Geschäftskosten und Gemeinkosten der Baustelle)	frais courants

cost comparison	Kostenvergleich, Bilanz	bilan des frais
failing	Mängel, Fehler	défauts
electronic computer	elektronische Rechenmaschine	calculatrice électronique
manual computation	Rechnen ohne Maschine	calcul sans machine
input data	Eingabewerte	données d'entrée
punch card	Lochkarte	carte perforée
output data	Ergebnisse	données de sortie
sorting, selecting	Sortieren, Aussortieren	trier
large memory	großer Speicher	grande mémoire
site record	Baustellenbericht	rapport de chantier
foreman's daily diary	(Polier-)Bautagebuch	registre de chantier
operation sheet	Arbeitsblatt	tableau d'avancement des opérations
to price	kalkulieren	calculer
feedback	Rückinformieren, Rückwirken, Rückkoppeln	réaction

Dialogue

Robert: Tell me, Duncan, what is your honest opinion about the necessity of a network programme?

Duncan: Well, in the case of a complex contract, or maybe one over a long period, I believe it's rather important. I must add, though, that it's of no earthly use, in my view, preparing a programme of such detail if it isn't properly utilised once the job is underway. It's rather like an overweight patient ignoring a diet sheet that a doctor has taken much trouble to prepare for his benefit.

Robert: In your experience then, have you often come across such wasted effort?

Duncan: Unfortunately, Robert, I have on a number of occasions. It's the poorly organised medium-sized contractor who usually comes unstuck on this score. The good intention is always there, but putting the theory into practice is quite another story.

Robert: You've probably found, as I certainly have, that the really large concerns, where the expertise is always at hand, can justify the time and expense of preparing elaborate pre-contract programming.

Duncan: Yes indeed, but you must remember, of course, that on massive contracts there's a lot of money to be gained by planning to the last detail, and a great deal more to be lost by failing to work rigidly to a plan. In fact, I know a number of large British firms who have even completed their contracts well before time, largely due to a high degree of critical planning.

Robert: I wouldn't for one moment detract from the obvious merits of network planning, but I've known contractors, including some large ones, to revert back to simpler systems, like the bar chart. Have British contractors, to your knowledge, done this?

Duncan: In the case of average-sized contractors, I would say this has often happened, simply because they don't have the necessary resources, experience, nor the ability of the larger contractors to properly implement a sophisticated programme.

Robert: So much for that topic, Duncan. Now I'd like you to explain, as briefly as possible please, the system, known as the line of balance method, for planning a contract. In fact, what does it entail and when is it used?

Duncan: First of all, I must confess I've had only limited experience of it so far, but I do know that on extensive housing schemes and similar repetitive projects it's beginning to make its mark here in Britain. I'll try to be brief in my explanation, but that may not be so easy.

Robert: Excuse the interruption before you get going, but why is it preferable to a network or a bar chart programme?

Duncan: I'm not going to stick my neck out by praising or critising it – it's quite adequate, to my mind, in the right situation – but I can assure you that it is a far simpler and more understandable programme for straightforward repetitive work.

Robert: In what way exactly?

Duncan: Well to begin with, it takes the form of a simple graph, usually showing the time factor along the bottom and the house numbers, for example, up the left hand side. Several curved or inclined lines, representing each operation, such as roof tiling or floor screeds, run up the graph approximately parallel to, yet never crossing each other.

Robert: From your description so far, I think I can picture what you mean. Now then, would I be right in saying that each horizontal line drawn from a house number on the left strikes the inclined lines or curves, and thus gives the dates on which each operation should start on any particular house?

Duncan: Yes, you've got the idea. You can appreciate, no doubt, that the information given is not very detailed, but on repetitive work this is hardly necessary.

Robert: That's perfectly true, because most contractors who specialise in housing are well versed by practical experience in how long each operation is likely to take and how many men will be required to execute the work. Incidentally, this so-called line of balance method is similar, if not the same in principle, to our Continental team work system for repetitive work.

Duncan: From what you've just said, I would suspect that the two systems only vary in name, but carry on please.

Robert: Each team of workers has a specific job to do on each building, and they work rather like a group of factory workers on an assembly line. The parallel I've drawn between these site teams and factory workers may not be strictly accurate, but it does, I'm sure, give the jist of the idea of logical repetitive working.

Duncan: Before we leave this topic. There's one question I'd like to put to you, and that is, does this team work system involve drawing up some form of graph or plan, or is it operated by rule of thumb?

Robert: A chart is normally prepared, but as you can imagine, it's not necessary to make it too elaborate. Provided it is fully understood by the teams and the planner, it serves its purpose in whatever form it takes. I don't think there exists any hard and fast rules.

Duncan: Having touched on the subject of assembly line type working, it seems opportune to mention an article I once read on speeding up building processes in Germany. This article spoke of much government backed research into possible methods of increasing productivity, and thus saving building costs. In Britain, we hear a great deal about increased productivity etc., and I'm always interested to know if the German government's efforts in this direction have, in fact, borne any fruit.

Robert: Yes, generally speaking, they have. A good example was in the development of a new town near Munich (München).

Duncan: You're referring to the overspill town called Perlach, are you not?

Robert: Yes that's the place all right, but please explain what you mean exactly by the term overspill town.

Duncan: That's easy. It is, as the name vaguely suggests, a new town, or even an existing town increased in area, to cater for the excess, or overspill in population and industry of a large town, which has already reached saturation point. In Britain, for example, Stevenage was one of the first new overspill towns intended mainly to ease the problem of London's ever increasing expansion. May I now ask you a question, Robert, on a very different subject?

Robert: Please do.

Duncan: On the Continent, how do you overcome the serious problem of rising costs and wage rates in respect of a contract? In other words, what provision is made in the contract documents for a contractor to recover these likely increased costs?

Robert: First of all, let me deal with short term contracts.

Duncan: Sorry to butt in, but what do you class as a short term contract?

Robert: I can't be absolutely certain, but I would say it's one for a period of up to about six months or so. Anyway, for such a contract, no clause is normally inserted in the contract conditions enabling the contractor to claim for increased costs during the contract period. Since any intended wage increases are known in advance for fixed periods, and the early bulk buying of materials is normal for a short term contract, no clause is necessary.

Duncan: Of course, the bulk buying of materials doesn't entirely eliminate the risk of overnight price increases, but most contractors, no doubt, would include some built-in contingency in the rates or prelims for the unpredictable

Robert: Now to turn to the long term contract. Naturally, things are somewhat different here.

Duncan: I can hardly imagine that Continental contractors are expected to tender for long term contracts on a fixed price basis in an inflationary economy, as is expected in Britain in many cases. As a matter of interest, the National Federation of Building Trade Employers (N.F.B.T.E.), a contractors' association in Britain, has been kicking against the government's policy of retaining fixed price tendering for some years, but as yet, with little success. This is, of course, an economic issue, but nevertheless, it does seem odd to me that industry in general is permitted to increase its prices to keep in step with the increasing cost of living, whilst contractors must normally accept firm prices. This puts the onus on them to include some allowance for likely increases.

Robert: As I see it, your system certainly leaves a lot to chance. Our system, I think, is somewhat better, in that the risk element is reduced in many long term contracts by the inclusion in the contract conditions of a special clause, literally translated as the "wage and material sliding clause".

Duncan: How does this clause reduce the risk?

Robert: Well, it does at least give the contractor a chance to price the bills of quantities at the current rates at the tendering stage, and to add an agreed, but limited percentage allowance under this clause for those unforeseen increased costs occurring during the contract period.

Duncan: This adjustment percentage on the contract figure is, I believe, agreed by all parties to the contract, is it not?

Robert: Yes, agreement must be reached at tender stage.

Duncan: Some British contract conditions do include a fluctuations clause, which is usually deleted, so most contractors are invariably forced to uplift their rates or prelims to allow for possible increases, or to insert an overall percentage adjustment for the same purpose. The whole thing, to my mind, is left far too much to guess work.

Robert: This is doubtless an interesting and relevant aspect of the planning and cost control of a contract, since increased costs do have a considerable bearing on both. Incidentally, contractors in Britain will find that Value Added Tax, yet another increased cost which Continental contractors have now learned to live with, has quite an effect on cost control, and cash flow thus becomes all the more important. However, all that is somewhat off the beaten track, perhaps, so I suggest we get a little closer to the main subject.

Duncan: Let's talk about sub-contractors, shall we?

Robert: By all means. The ever increasing tendancy for contractors to utilise sub-contractors is, I'm sure, becoming a significant factor which influences planning and costing. This topic is really very much in your line, I know, so maybe you should begin.

Duncan: I'm delighted, Robert, that you've brought up the question of sub-contracting, and I hope you won't regret having done so. It's a pet subject of mine, on which I have very decided views, as you well know. If I ramble on too much, please stop me.

Robert: Well, I know and sympathise with many of your views, Duncan, and I'm sure a lot of our readers will benefit from your personal experiences in this field.

Duncan: To begin with, I sometimes feel that sub-contracting has got rather out of hand, particularly in Britain in recent years. One can even find cases here of sub-contractors sub-letting their work to other sub-contractors.

Robert: In what way exactly do you think the situation is now out of hand?

Duncan: The fact that most of the work on contracts is done by sub-contractors these days, and not by the main contractors themselves, as was the case some years ago, is in my opinion a retrograde step. Many main contractors tend to become agencies for outside labour, and to a large extent, their only real concern lies in the financing, planning and controlling of the contract.

Robert: Well, when you tell me that even one man doing odd jobs on a contract – tea making included – can be a self-employed labour-only sub-contractor, I can fully understand your strong views on the subject. You are suggesting, in fact, and please correct me if I have the wrong impression, that main contractors in many cases are only associated with a contract in name alone.

Duncan: Apart from the smaller contractors who do mainly jobbing and minor alteration work, I firmly believe that my views are well founded. This is certainly true in Britain. I've spoken to a number of contractors, large and even small, who admit that they only actually perform about 10 % of the site work themselves. My own experience has confirmed that this is perfectly true.

Robert: Before you go further ahead, tell me why you consider such a situation has damaged the construction industry.

Duncan: This is simply answered by the fact that far less control is possible over sub-contract labour than directly employed labour. The main reason being, that sub-contractors are very independent and often unreliable.

Robert: That's fair comment. It's plain to see that sub-contractors, who now constitute such a large slice of the main contractor's resources, can so easily disrupt any network or other form of plan, and this makes planning far more difficult.

Duncan: You've hit the nail right on the head, Robert. I would say though, that the larger contractors do have more control over their "subbies" than smaller contractors do.

Robert: This is so, I presume, mainly because large contractors can offer better incentives, more work and a reasonable guarantee of continuity of work to their "subbies", as you call them.

Duncan: Yes, this is true. You see, most sub-contractors prefer to work on the larger contracts for the obvious reasons you've just mentioned, and it therefore means that large contractors can often afford to pick and choose a little from several competing sub-contractors. In this respect, the smaller contractors don't usually have the whip hand, and so the sub-contractors all to often call the tune.

Robert: Don't you think that main contractors generally are largely to blame for the present situation, because they themselves have encouraged sub-contracting, in order to cut down their overheads in the hope of increased profits?

Duncan: Up to a point this is true, since they don't pay any on-costs on indirectly employed labour, and have fewer directly employed men to find work for in lean times.

Robert: Do you really think though, that many contractors stand to gain much by using so many "subbies"?

Duncan: Apart from the large well organised concerns, I don't think so, because they tend to overlook the hidden costs of arranging, controlling, measuring-up and paying the "subbies". Further costs often arise when bad work has to be put right, usually by the contractor's own limited labour force. It should be said, however, that sub-contracting is also partly caused by the natural desire for workers, like anyone else, to be their own bosses.

Robert: The last point you made is a very real one, the importance of which is rarely highlighted.

Duncan: On the Continent, of course, the Common Market does help to provide a free flow of labour from one European country to another, and this must ease your contractors' labour problems to a certain extent.

Robert: That's true. Now I realise that I'm encouraging you to labour your viewpoint on the disadvantages of sub-contracting, but perhaps you should explain the expression, "lump", to which I have so often heard you refer. Go ahead, Duncan, I've left the door wide open for you.

Duncan: I need little encouragement, I can assure you Robert, on this particular topic. The frequent use of the "lump", or labour-only gangs, in Britain is a sore point with me. Having had to deal with this form of self-employed sub-contractor, including even one-man bands, on many occasions, I do know the problems involved. Mind you, I have come across extremely good gangs, but the quality of some of the work turned out is, in my opinion, pretty ropey.

Robert: Surely, site supervision is chiefly at fault in such cases.

Duncan: Yes, up to a point I agree, but neither the foreman on site nor the contract manager has the same degree of control over the "lump" as he has over his own men. The constant threat of any of these very independent gangs or individuals failing to complete their work can, and sometimes does land a contract in queer street. In other words, they are virtually able to hold a main contractor to ransom. I must say, in all fairness, that efforts have been made for some time to reduce the use of the "lump". The Income Tax authority and trade unions continue to exert some pressure, and recent unemployment too has had its effects.

Robert: This problem of sub-contracting also exists on the Continent, but I'm sure it isn't so acute here as it appears to be in Britain. We have sub-contract gangs for specialised work, but very small groups or one-man bands are rare. It seems to me that you have suffered a greater percentage drop than we have in recent years in the number of workers directly employed in the construction industry. I believe there are many reasons for this, including economic ones, but to go into those now would lead us once again far from our subject.

Duncan: Then I think we should get back on course.

Robert: You mentioned the foreman a moment or so ago, and here we have, to my mind, one of the most under-rated members of the contractor's team when it comes to implementing the plan.

Duncan: I couldn't agree with you more. Here again, I have had first hand experience, and you can take it from me that they differ from chalk to cheese. A good foreman is worth his weight in gold, but a poor one spells disaster. This may all sound rather dramatic, but he is, after all, a key figure linking the working force on site with the administration staff, who are often too remote from the scene. He is in very close contact with the operatives and the actual work in progress, and is thus the first to know when things aren't going according to plan.

Robert: His ability and alertness, in fact, can do a lot to make a job really tick. The practical experience of a foreman also proves useful in future contract planning.

Duncan: Well, I hope our few words about the foreman has put him on the map a little, and on that note I feel we ought to leave this particular aspect of construction, on which there are so many divided opinions.

Vocabulary	Wörterverzeichnis	Glossaire
to be of no earthly use, (to be futile)	keinen praktischen Wert haben	être sans utilité
diet sheet	Diätplan	tableau diététique
to come unstuck, (to fail)	scheitern	échouer
on this score, (in this respect)	in dieser Hinsicht	de ce point de vue
expertise, (skilled and competent staff and organisation)	fachlicher „Apparat"	experts, groupe de spécialistes
pre-contract programming	Arbeitsvorbereitung	préparation du travail
critical planning	detaillierte Vorplanung	étude détaillée du projet
line of balance	Linienplan	directives d'équilibre
housing scheme	Wohnungsbauprojekt	projet de construction de logements
to make its mark, (to become popular)	Eindruck machen, sich einen Namen machen	faire impression, devenir populaire
to get going, (to begin)	anfangen, „loslegen"	démarrer
to stick one's neck out, (to commit oneself, to make a bold statement)	kühne Behauptung aufstellen	émettre des prétentions audacieuses
graph	graphische Darstellung, Diagramm	diagramme
roof tiling	Dachziegeleindeckung	couverture en tuiles
floor screed	Estrich	revêtement de sol
to picture	sich vorstellen, ausmalen	se représenter
to be well versed in, (to be familiar with)	bewandert (versiert) sein mit	être familiarisé avec
team work	Taktarbeit, Gruppenarbeit	team work, travail d'équipe
team, group	Gruppe, Kolonne	groupe
assembly line	Fließband	chaîne de montage
the jist of the idea, (some idea)	ungefährer Begriff	idée approximative
by rule of thumb (by past experience)	nach Daumenregel, erfahrungsgemäß	à vue de nez, empirique
government backed	von der Regierung unterstützt (gefördert)	promu par le gouvernement

to bear fruit	Früchte tragen	porter fruit
overspill town	Satellitenstadt	ville dortoir
wage rate	Lohnsatz, Tarif	tarif
to butt in, (to interrupt)	unterbrechen	interrompre
short term contract	kurzfristiger Vertrag	contrat de courte durée
clause	Klausel	clause
wage increase	Lohnerhöhung	augmentation des salaires
bulk buying	Großeinkauf	achat en gros
contingency, allowance	Reserve (betrag)	participation, allocation
built-in contingency	stille Reserve	réserve cachée
prelims, preliminary items	Vorbemerkungen	préliminaires
long term contract	langfristiger Vertrag	contrat de longue durée
fixed price	Festpreis	prix fixe et définitif
National Federation of Building Trade Employers, N.F.B.T.E., (British association to protect the interests of contractors)	Unternehmerverband der britischen Bauindustrie	fédération nationale des entrepreneurs britanniques de la construction
to kick against, (to oppose)	entgegentreten	s'opposer à
economic issue	wirtschaftliche Angelegenheit	résultat économique
fixed price tendering	Kostenanschlag zu Festpreisen	offre à prix fixe
to keep in step	Schritt halten	aller de pair
to leave to chance	dem Zufall überlassen	laisser au hasard
material and wage sliding clause	Stoff- und Lohngleitklausel	clause sur l'échelle mobile des salaire et des prix
current rate	Tagespreis	prix courant
tendering stage, tender stage	Angebotsstadium, Zeitpunkt des Angebots	stade de l'offre
percentage allowance	prozentuale Preisänderung	pourcentage autorisé
adjustment percentage	(Preis-)Anpassungs-Prozentsatz	pourcentage d'ajustement
contract figure	Auftragssumme	montant du marché
party	(Vertrags-)Partei	partie
fluctuations clause	Gleitklausel	clause de fluctuation
overall percentage adjustment	allgemeine prozentuale Anpassung	pourcentage général d'ajustement
guess work	Mutmaßung, Vermutung	hypothèse
Value Added Tax (V.A.T.)	Mehrwertsteuer	TVA
to be off the beaten track, (to be off the subject)	vom Thema abweichen	sorti du sujet
to be in someone's line, (to be someone's special interest)	auf jemandes Linie liegen, zu jemandes Interessengebiet zählen	être dans la ligne de quelqu'un
pet subject (favourite subject)	Lieblingsthema	thème préféré
to ramble on, (to go on talking for a long time)	sich in Reden verlieren	discúter sans fin
to get out of hand, (to become uncontrollable)	den Händen entgleiten	perdre le contrôle de
to sub-let (work), (to pass work to somebody else to do)	(Arbeit) untervergeben	donner en sous-ordre
retrograde step	Rückschritt	rétrograder
outside labour	firmenfremde Arbeit(er)	travail extérieur
odd jobs	(gelegentliche) kleine Arbeiten, Ausbesserungs- und Reparaturarbeiten	petits travaux

self-employed labour-only sub-contractor	selbständiger Subunternehmer für Bauarbeiten (ohne Materiallieferung)	sous-traitant indépendant fournissant seulement sa main d' œuvre
to be associated in name alone	nur dem Namen nach verbunden sein	associé nominalement seulement
jobbing, jobbing work	Akkordarbeit	travail aux pièces, à la tâche
alteration work	Umbauarbeiten	travail de modification
"subby", sub-contractor	Subunternehmer, Nachunternehmer	sous-traitant
to pick and choose, (to be selective)	(die besten) „herauspicken" und auswählen	choisier le meilleur
to compete	konkurrieren, wetteifern	concurrencer
to have the whip hand, (to be in control)	die Zügel in der Hand haben	avoir les brides bien en main
to call the tune, (to dictate the terms)	den Ton angeben	donner le ton
lean times, (times of little prosperity)	auftrags- bzw. arbeitsschwache Zeiten	périodes de faible emploi
to stand to gain	sicherlich gewinnen, ein Gewinn zu gewärtigen haben	être sûr d'être bénéficiaire
measuring-up	Aufmessen	métrer
to be highlighted, (to be emphasized)	ausdrücklich betont (hervorgehoben) werden	être souligné
Common Market	Gemeinsamer Markt	Marché Commun
European Economic Community (EEC)	Europäische Gemeinschaft (EG)	Communauté économique européenne
to labour one's viewpoint, (to overstress one's opinion)	jemandes (eigene) Meinung „strapazieren"	exagérer l'opinion de quelqu'un
"lump", labour-only gangs	selbständige Kolonne für Bauarbeiten im Werkvertrag ohne Materiallieferung	équipe de travail indépendante et foraine
sore point	wunder Punkt	point sensible
one-man band	‚Ein-Mann-Kapelle', ‚Gruppe' aus einem Mann	équipe d'un seul homme
pretty ropey, (rather poor)	ziemlich minderwertig	plutôt maigre
site supervision	Beaufsichtigung der Arbeit auf der Baustelle	surveillance du travail sur le chantier
to be at fault	verfehlt sein	être manqué
to land someone in queer street, (to put someone in difficulty)	jemanden in Schwierigkeiten bringen	mettre quelqu'un en difficulté
to hold someone to ransom, (to make someone do as one wishes)	jemanden seinen Willen aufzwingen	faire entendre raison, imposer sa volonté
Income Tax authority	Einkommensteuer-Behörde	bureau fiscal de l'impôt sur le revenu
trade union	Gewerkschaft	syndicat
percentage drop	prozentuale Abnahme	pourcentage de dégression
underrated	unterbezahlt	sous-payé
first hand experience	Erfahrung aus erster Hand	expérience directe
to differ from chalk to cheese, (to differ very much)	sich unterscheiden wie Tag und Nacht	être totalement différents, se distinguer comme le jour et la nuit
to spell disaster, (to bring disaster)	Unheil anrichten, Katastrophe verursachen	amener le désastre
key figure	Schlüsselfigur	chiffre clé
to make a job tick, (to make a job run well)	eine Arbeit voranbringen	faire marcher un travail

to put someone on the map, **(to give someone some status)**	jemanden ins richtige Licht rücken, ihm Ansehen verschaffen	mettre quelqu'un en vue

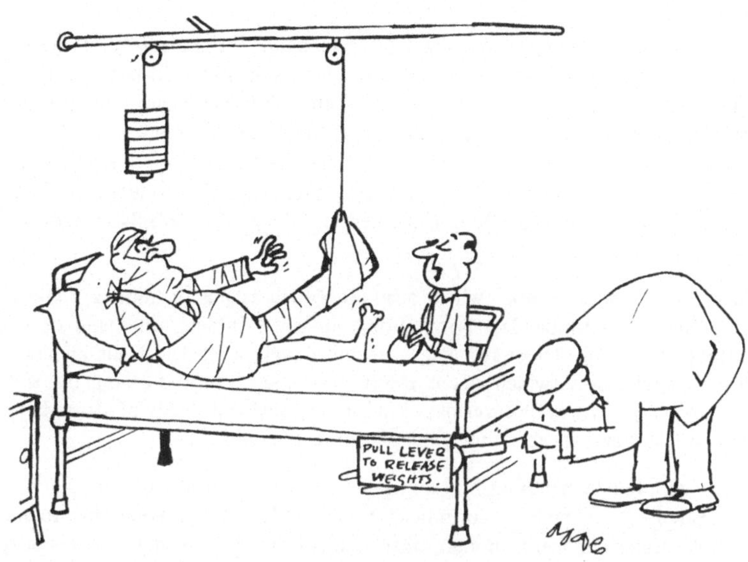

"Fred wanted to come along to apologise for tampering with the scaffolding."
(Reproduced by permission Construction Surveyors' Institute)

10. SAFETY ON CONSTRUCTION SITES

There are literally hundreds of enforceable regulations in being for the safety of site personnel and the general public, and also to safeguard property. In spite of considerable efforts to enforce safety measures, which are generally well publicised and displayed on sites, the injury and death rate, not to mention damage by fire etc., is still far too high. Safety precautions are basically common sense and a little consideration for others. Conforming to simple precautions not only preserves life and limb, but saves time and unnecessary expense in the long run. A site administration conscious of safety usually produces an efficient and tidy site, on which everybody knows where everything is located and dangerous hazards are reduced to a minimum, if not eliminated.

It's the small hazards and booby traps which account for thousands of injuries, some fatal, every year. Protruding nails in pieces of timber left lying about, unlashed ladders, unlit and obstructed staircases, overhanging scaffold poles and planks at head level and dangling electric leads at ankle level are some typical examples of carelessness, which often result in serious injury. Leaving lift shaft and manhole openings without adequate barriers or covering, or failure to provide a sturdy platform between a passenger hoist and a staging or floor level are recurring examples of death traps.

The methods often adopted in handling building materials and disposing of debris and surplus materials leave a lot to be desired. Throwing bricks or blocks, for example, from one operative to another during the offloading of a lorry seems innocent enough, until one is dropped, causing minor injury and irritation. Lifting heavy concrete units or steel members, which are insecurely fastened, with a block and tackle may appear safe, but should the load come adrift, any operatives at ground level would be lucky to escape disaster.

The disposal of debris from demolitions requires particular care and attention for a number of reasons. Quite often it has to be lowered from a considerable height into lorries or waste containers, and if this process is not strictly controlled and a proper chute used, broken bricks and other sharp pieces of debris are

Find out the two meanings!

(Reproduced by permission. British Safety Council)

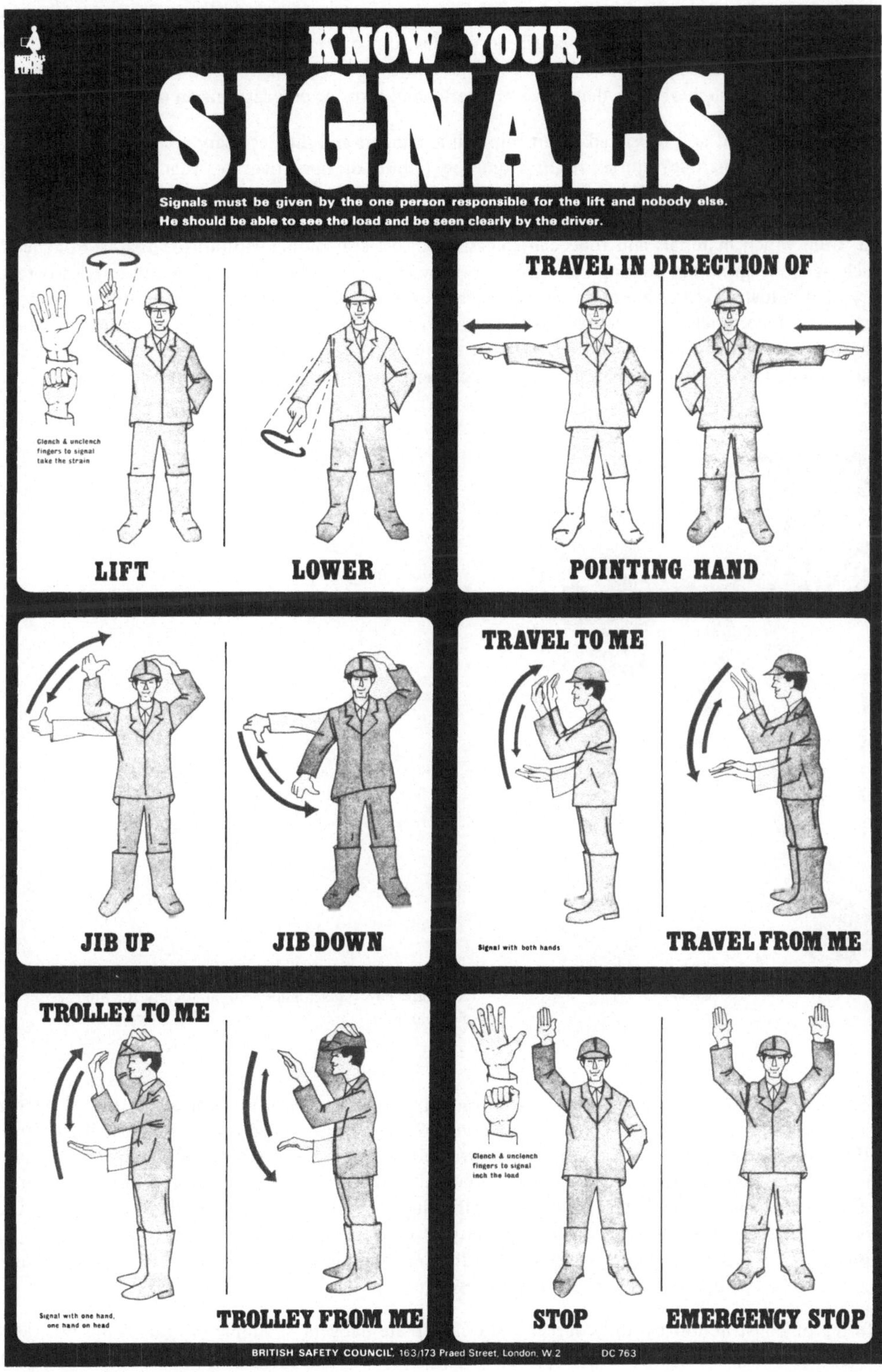

(Reproduced by permission. British Safety Council)

likely to inflict horrifying injuries. Such work inevitably produces much dust and flying grit, so those workmen in close attendance should at least wear goggles to protect their eyes. Industrial face masks may also be advisable in instances where dust could cause lung infection. The loading of debris too calls for some thought, so as to avoid such obvious dangers as an overloaded lorry or one carrying an overhanging load.

A great deal can be said and legislated about impending dangers and the necessary precautions to be taken against them, but numerous injuries result from the failure of operatives to wear protective clothing provided for their own safety. Helmets and steel-capped boots should always be the standard head - and footwear of all those engaged on a construction site. This is particularly important where heights are involved, from which materials and tools can so easily fall off scaffolding without toe boards. For specific jobs, such as stone grinding, chasing concrete or brickwork and welding, the wearing of goggles is really necessary, and industrial gloves are often recommended for handling rough materials, like ordinary concrete blocks. Some sub-contractors are rather loathed to wear any protective clothing at all, simply because it might slow down their production and reduce their earnings. This viewpoint is a little short-sighted, since they will often be faced with a hard legal battle for compensation in a case of serious industrial injury.

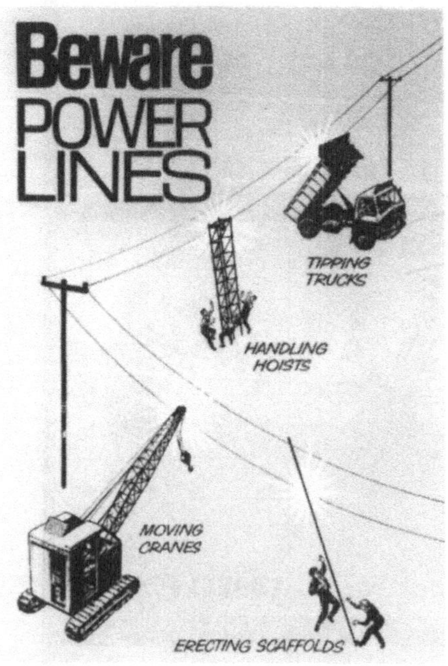

(Reproduced by permission. Royal Society for the Prevention of Accidents)

Safe working is too often sacrificed for increased productivity, especially by self-employed sub-contractors. Bonus and incentive schemes for the contractor's own workmen can be a real threat to safety, if the targets set by them are unrealistic. Every day, many unwarranted risks are taken, and although the majority evade any consequences, there is always someone who eventually suffers or causes suffering through overconfidence. During the erection and dismantling of scaffolding in particular, alarming risks often take place. Steel erectors and roof tilers too are sometimes guilty of carelessness at precarious heights and in slippery conditions in wet weather. A good saying, well worth remembering when one is tempted to take a chance, is "it's better to be Mr. Brown late than the late Mr. Brown".

Some forms of construction by their very nature can create dangers, if simple safeguards are not taken. Projecting reinforcing bars, like starter bars for columns etc., is one particular example of a potential hazard, where they are not capped off. In certain locations, open excavation without danger lamps at night is another common example. Mechanical plant, in general use on most sites nowadays, accounts for quite a number of fatal accidents. This is largely due to reckless handling or the lack of guards over the moving parts of such items as winches.

Minimising the consequences of accidents is as important as taking precautions against them. Adequate first-aid facilities, which are easily accessible on site, are therefore essential, and bandages, dressings, plasters, antiseptics etc., should always be regularly restocked. A shortage of these at the crucial moment could jeopardise any chance of stemming the ill-effects of injury, or even of saving life. It is desirable to ensure that all site personnel have received basic training in first aid, since a doctor cannot always be at hand when required. Preventing the spread of fire also calls for some basic training of personnel. Everyone on site ought to be familiar with the correct operation of the particular fire-fighting equipment, otherwise valuable time will be lost and most of the damage done long before the fire brigade arrives. Full water buckets located at various accessible points are also useful at such times of emergency.

To achieve a high standard of safety, a good example must be set by the foremen, contract manager and other site staff. A site with well sign-posted access roads, properly arranged storage areas in the compound etc., helps enormously to reduce many of the unnecessary hazards. Regular site visits by a fully trained safety officer, and not just anyone assigned to the job, are both essential and benefical on any contract.

Vocabulary	Wörterverzeichnis	Glossaire
safety	Sicherheit	sécurité, sûreté
construction site	Baustelle	chantier de construction
regulation	Vorschrift	prescription
site personnel	Baustellenpersonal	personnel de chantier
general public	Allgemeinheit	communauté
to safeguard	schützen	protéger
safety measure, safety precaution	Sicherheitsmaßnahme	mesures de sécurité
injury and death rate	Unfall- und Todesrate	nombre de blessés et tués par accident
damage by fire, fire damage	Brandschaden	dégâts d'incendie
life and limb	Leib und Leben	vie
site administration	Bauleitung	direction des travaux
hazard	Gefahrenrisiko	risque
booby trap	versteckte Gefahrenquelle	source de danger cachée
protruding nail	herausstehender Nagel	clou sortant
unlashed ladder	ungesicherte Leiter	échelle non attachée
staircase	Treppenhaus, Treppe	cage d'escalier
scaffold pole	Gerüstriegel	traverse d'échafaudage
scaffold plank	Gerüstbohle	madrier d'échafaudage
head level	Kopfhöhe	à hauteur de la tête
electric lead, electric supply cable	elektrische Leitung, ~Kabel	câble électrique
ankle level	Fußknöchelhöhe	à hauteur de la cheville
lift shaft	Aufzugschacht	cage d'ascenseur
manhole opening	Schachteinstieg -Öffnung	ouverture de regard
barrier	Absperrung	barrière
covering	Abdeckung	couvercle
platform	Bühne, Podest	plate-forme
passenger hoist, lift	Personenaufzug, Fahrstuhl	ascenseur
staging, stage	Zwischenpodest	palier intermédiaire
floor level, floor height	Stockwerkhöhe, Etagenniveau	hauteur d'étage, niveau d'étage
death trap	Todesfalle	piège mortel
to dispose	wegschaffen, beseitigen	disposer
debris	Trümmer	débris
surplus material	übrigbleibendes Material	matériel excédentaire
brick	Mauerziegel	brique
block	(meist großformatiger) Beton-, Hohlblockstein	bloc, bloc creux

operative	Arbeiter	ouvrier
offloading, unloading	Abladen	décharger, déchargement
lorry	Lastwagen	camion
concrete unit	Beton(fertig)teil	élément préfabriqué en béton
steel member	Stahl(bau)teil	élément en acier
block and tackle	Flaschenzug	palan
at ground level	auf Erdgleiche, auf dem Boden	au niveau du sol
demolitions (usually plural)	Abbruch	démolition
waste container, disposal unit	Müllbehälter	récipient à ordures
chute	Rutsche	goulotte
grit	Grus, Grobstaub	gravillon
goggles	Schutzbrille	lunettes protectrices
face mask	Staubmaske	masque à poussière
protective clothing	Schutzkleidung	vêtements de protection
helmet	Helm	casque
steel-capped boot	Stiefel mit Stahlkappe	souliers à bout protecteur en aci
scaffolding, scaffold	Baugerüst	échafaudage
toe board	Bordbrett	planche de bordure
to grind	schleifen	gréser, meuler
to chase concrete, (to cut a channel or sinking in concrete)	Schlitze o.ä. in Beton stemmen	creuser dans le béton
to weld	schweißen	souder
industrial gloves	Arbeitshandschuhe	gants de travail
sub-contractor	Subunternehmer, Nachunternehmer	sous-traitant
legal battle	Rechtsstreit	conflit juridique
compensation	Schadenersatz, Vergütung	indemnité
self-employed sub-contractor	mitarbeitender Subunternehmer	sous-traitant travaillant lui-mêm
bonus	Prämie	bonus
incentive scheme	Zulagenplan für Mehr- und Sonderarbeiten	tableau des suppléments pour travaux supplémentaires ou spéciaux
target	Akkordsatz, Vorgabeleistung	taux du travail à la tâche
to dismantle	abbauen	démanteler
steel erector	Stahlbau-Monteur	monteur de construction en acie
roof tiler	Dachdecker	couvreur
reinforcing bar	Betonstahlstab	barre de renforcement
starter bar	Anschlußeisen	fer de raccordement
column	Säule, Pfeiler	colonne
to cap off	mit Kappe versehen	décaper
excavation	Aushub	creuser
open excavation	offene Baugrube	fouille ouverte
danger lamp	Warnlampe, Absperrungsleuchte	signal lumineux
mechanical plant	Baumaschinen und -geräte	engins et machines de constructi
guard	Schutzvorrichtung, (Wache)	protection, (garde)
winch	Winde	treuil
first-aid facilities	Einrichtungen für erste Hilfe	équipements de premier secours
bandage	Binde	bandage
dressing	Verband, Verbandstoff	panser
plaster	Heftpflaster	sparadrap
antiseptic	keimtötendes Mittel	antiseptique
to restock	wiederbeschaffen	réapprovisionner
basic training	Grundausbildung	formation initiale
fire-fighting equipment	Feuerlösch-Einrichtung	installation contre l'incendie
fire brigade	Feuerwehr	service de pompiers
foreman	Polier	chef d'équipe

contract manager, agent	Bauführer (des Bauunternehmens)	chef de chantier
site staff	Baustellenpersonal	personnel de chantier
sign-posted	beschildert	signalisé
access road	Zufahrtsweg	voie d'accès
storage area	Lagerplatz	aire de stockage
compound	umzäuntes Gelände	terrain clôturé
safety officer	Sicherheitsbeauftragter	chargé de la sécurité
contract	(Bau-)Auftrag	ordre

Dialogue

Robert: When thinking of safety matters, I'm often reminded of my very first experience of a serious accident on a construction site, and it has certainly left a marked impression on me. I can recall passing a long access tunnel, which led to an excavated pit for the extension of an underground project, when I was met by two men carrying a badly injured young carpenter on a stretcher. The poor chap had to have his leg amputated several days later.

Duncan: What was the actual cause of that accident, Robert?

Robert: Well, to put it bluntly, it was sheer carelessness. Apparently, a group of workmen had been instructed to lift two heavy steel beams – soldier beams, as a matter of fact – using a steel hoisting cable. The cable snapped, and the "chippy", who was unfortunate enough to be close by, was pinned to the ground by one of the falling beams.

Duncan: Can you remember who was held responsible for the accident?

Robert: Yes, Duncan, it was the foreman who carried the can, since he was in charge of the operation, and that included the responsibility for the safety of his operatives. In this particualar incident, he had permitted the use of a cable which wasn't man enough to take the load.

Duncan: No doubt he was under pressure to get a move on, which is so often the case, and didn't think to check the cable. This sort of thing happens time and time again, especially where the contract is running behind schedule or the men are on a bonus scheme.

Robert: Actually, there was an incentive scheme on that contract, come to think of it.

Duncan: Of course, extra effort should always be rewarded by extra payment – nobody in their right mind would dispute that – but one has to draw the line somewhere in the interests of safety.

Robert: The danger creeps in if men get slaphappy. This could have been the case in the handling of those beams. I could well imagine the temptation to loop the cable around more than one girder at a time, to save time and because it's easy to do so, since the flanges tend to cling together.

Duncan: There's certainly a lesson to be learnt from the case we've just discussed. Another point worth talking about is the common failure of some contractors to give their plant and equipment a regular inspection for wear and tear.

Robert: Tell me, Duncan, how often is your plant in Britain inspected by law, and who carries out the inspection? Before you answer, I'd also like to ask you which of your machines in particular are subjected to such an official examination?

Duncan: Quite a barrage of questions, I must say. By law, all our plant must have an insurance inspection every 14 months, and this is carried out by the Government Factory Inspectorate. In addition to this, a weekly inspection should also be made by a competent person on site, who should issue a certificate to the site office. Now to answer your third question. Cranes and other lifting equipment are the items which receive the most severe tests in Britain, and this isn't surprising.

Robert: That's quite interesting. I've got another question for you on British practice. Do your operators of excavators and tower cranes, for example, require a special license as an additional safeguard?

Duncan: No, there isn't any special licensing system in Britain for operators of contractor's equipment. All that is necessary, oddly enough, is a normal driving license when a machine is being driven on the public highway.

Robert: With regard to the inspection of plant, I can assure you that the situation here on the Continent is quite similar to that in Britain. In Germany for example, tower cranes have to be thoroughly inspected every 4 years by the so-called T. Ü. V., an association authorised by the Government. Contractors still have their responsibilities to check building machines weekly and to ensure that operators are properly trained and hold the necessary licenses.

Duncan: On the subject of plant, it's a debatable point whether hired plant is maintained up to a higher safety level than that owned by the contractor himself. As you know, plant hire in Britain, even by large concerns, is very common, far more so than on the Continent. What are your views on this aspect of safety?

Robert: With smaller firms, I would think that a far higher safety standard was generally achieved with hired plant, because plant hire firms have special workshop facilities and trained mechanics for proper maintenance and repair.

Duncan: I'm sure you're right, Robert, since they usually carry enough plant to justify quite an elaborate set-up. Another point, of course, is that plant hire firms don't usually run their machines into the ground, as do most small contractors who have their own.

Robert: What you're saying, in fact, is that plant hire firms normally have newer machines etc., or at least reconditioned ones. This in itself ensures a better safety standard.

Duncan: Turning now to first-aid training, which is closely linked with safety, I have to admit rather shamefully that scarcely anyone in Britain's building industry is obliged to take a course. In fact, this is so in nearly all walks of life in Britain.

Robert: As a Continental, I find that very surprising. In Germany, for example, things are not quite so lax, as you know.

Duncan: I believe I'm correct in saying that Germans can't get a new driving license until they pass a first-aid test.

Robert: That's perfectly true, and I think it makes good sense. Since most working people in Germany hold a normal driving license, this obviously reduces the number of site operatives who must have special first-aid instruction.

Duncan: Another important thing relating to safety is taking the necessary precautions against electrical dangers. First and foremost, an electrical installation must be perfectly safe, whether it forms part of the permanent structure or is for temporary site lighting and power. As you are more electrically minded than I am, you will, no doubt, want to enlarge on what I've just said.

Robert: Yes indeed, but first of all, let's get to the root of the matter. A safe installation basically means correct wiring up, proper insulation of cables and such connections as plugs and junction boxes, and last but not least, the careful positioning of cables etc., so that they don't come into contact with working plant, or operatives for that matter. This may sound like a tall order on a site bubbling over with activity, but it must be borne in mind that electricity, when mishandled, is really lethal. Like fire or water, it's a good servant but a bad master. Incidentally, Duncan, do you reckon Britain has yet reached a satisfactory safety standard in this respect?

Duncan: Well, I certainly believe we are making some progress in that direction, but we've a heck of a long way to go yet. Take the simple plug as an example.

Robert: Sorry to interrupt you, but I understand that the problem of introducing satisfactory plugs, both from the safety and handling point of view, is common to Britain and the Continent. At the moment, we have at least three different types of plug here on the Continent. However, efforts are being made to bring in a standardised plug, known as the European plug.

Duncan: I can see our discussion is gradually moving towards the subject of electricity – one of your favourites, I know – but I'm sure most readers will appreciate its significance in relation to safety. Would you mind carrying on, Robert, where you left off

Robert: Before mentioning other electrical topics, I'd like to continue discussing plugs for just a little longer, if I may.

Duncan: Sure, but firstly would you say something about them in respect of the C. E. E. system – that is the system introduced by the International Commision on Rules for the approval of electrical equipment.

Robert: Certainly. According to this so-called system, Continental manufacturers of new 3-phase electrical appliances, for example, are obliged to provide them with the safer round shaped 5-point non-interchangeable plug. You know this plug, of course, but what do you think of it?

Duncan: To my mind, it's very neat and sensible.

Robert: What are the chances of seeing it universally employed in Britain?

Duncan: I can only say that one always hopes for this kind of standardisation of plugs, amongst other items, in Britain, where we still have a number of different systems. Various costly changes have already taken place here recently to standardise and improve our electrical systems, and to make further sweeping changes at the moment to the European system would probably meet with some resistance for economic reasons.

Robert: With regard to your various systems, which you just mentioned, I'm sure it will be of some interest to the Continental, if you could give one or two examples of the different plugs and socket outlets. I found the variety quite intriguing, but not always in the best interests of safety, to be honest. To be fair, though, this variety mainly occurs in older property.

Duncan: Well, let's just consider the single-phase domestic plug. These vary both in size and shape, depending on whether they are 15 amp, 13 amp or 5 amp, and to add to the confusion, the 5 amp type often varies in itself, depending on whether it is 3- or 2-pin. To complicate matters still further, it's possible even nowadays to get 3-square pin and 3-round pin plugs, although it's only fair to say that most plugs being manufactured today are predominantly large 13 amp 3-square pin.

Robert: Before you go on, I must chip in to mention that the earth terminal pin of most of your 3-pin plugs is larger in section than the two live pins. I find this a little odd.

Duncan: That's perfectly true. At this point, it would be as well to briefly describe the socket outlet into which this type of plug engages. The contact holes to receive the live pins are usually closed by a shutter, and it can only be opened by the advanced entry of the larger earth pin. This is purely a safety precaution to stop children from poking a piece of wire, or something similar, into the live contacts of the socket outlet when the plug is withdrawn.

Robert: A feature of most of your plugs is the inclusion of a fuse, which is bound to make it more bulky.

Duncan: This may be so, but a fused plug has its advantage. In the event of an electrical fault in an appliance, only the plug fuse is blown and no other electrical appliances on the same circuit are temporarily put out of action.

Robert: Just another word or two now about your socket outlets. Many of these, I've noticed, are provided with switches. This isn't so here on the Continent.

Duncan: Yes, you're right. Switch socket outlets are common in Britain, but I won't go into the "pros and cons" now. Maybe we like them, because we're a little overcautious in our means of disconnecting appliances.

Robert: On the whole, there are marked differences in our systems, to speak of plugs alone. The most common Continental plug simply has two small round pins for the live contacts and employs a scraping earth contact on its outside edge. Incidentally, we never have fused plugs nowadays.

Duncan: I don't think we need say any more on plugs and socket outlets. In fact, I suggest we broaden our discussion a little, even if we do keep to electricity.

Robert: Let's discuss voltage in relation to safety, if you like.

Duncan: A good idea, but I think we must begin by outlining some of the basic differences in our two voltage systems. Perhaps you could take it up from there, Robert.

Robert: Yes I'd like to. I understand that the British network supplies 415/240 volts in most areas, as opposed to a power supply of 380/220 volts almost universally adopted throughout the Continent.

Duncan: Generally speaking, I believe you're right. In my home city, however, the Electricity Board's supply is at present 398/230 volts in most districts, but I think the eventual intention all over the country is to change over to 415/240 volts.

Robert: The fact that there seems no likelyhood in the immediate future of a common European voltage system is a pity, but maybe we should be thankful that we do at least have a common frequency cycle of 50. I should add though, that a slightly higher frequency, such as 60 cycles employed for example in the U. S. A., Brazil 'and Japan, is more economical.

Duncan: Now I suggest we relate voltage to safety. Tell me, do you supply a lower voltage in special cases, as we do?

Robert: Yes we do. In the case of an electrically driven concrete poker for example, only 42 volts are permitted. For other electric tools which are portable, a completely insulated casing is required if they are not specifically designed for low voltage – that is 42 volts.

Duncan: We have similar regulations for certain cases. For site work, it is strongly recommended that 110 volts be used through a special transformer with its central point earthed.

Robert: With such a system, you see, those operatives in direct contact with the earth whilst handling power tools, portable hand lamps etc., can only be subjected to 55 volts, if the worst comes to the worst. That's a rather clever and safe system, don't you think?

Duncan: I agree, as long as a really good earth is provided. You know as well as I do, the sort of earth that is so commonly used. A steel rod of insufficient length or thickness driven just a little way into the ground, or an earth clamp with an inadequate metallic contact are a couple of classic examples of poor earthing methods. Incidentally, on the subject of earthing, it's worth mentioning that in Britain, the Institute of Electrical Engineers' Regulations require the installation of an earth-leakage circuit breaker on all alternating current circuits on sites where the power supply exceeds 125 volts. This is in addition to protection by fuses.

Robert: This is a good point, I feel, at which we could well turn our attention to the risks involved with electricity, the effects of shock and some of the possible means of avoiding trouble. To start the ball rolling, let's begin with the risk element. The degree of risk associated with electric shock through alternating current depends, as you know, on the value of the current, its frequency, its duration and the path it takes through the human body.

Duncan: The path, of course, can differ according to the point of contact. I mean, for example, that contact could be through a tightly gripped hand, unable to relax.

Robert: Yes, your comments are quite relevant. Now to try to explain a little further. This value of current depends on the voltage applied and the overall resistance – that is the resistance of the contacted skin area and the way through the body to the other point, which is usually the earth.

Duncan: As we are all well aware, there are many ways in which a site operative can get an electric shock, but the most common and alarming cause probably stems from handling the metal casing of defective electrical equipment. Proper maintenance would get rid of that cause straightaway.

Robert: It's bad enough getting a shock of any kind, but the effects of some can be very serious. Take for example a sudden shock, where the stimulation of the nerve ends is conveyed to the central nervous system.

Duncan: Very unpleasant indeed. An even more serious situation arises when fibrillation of the heart occurs. One instance where this can happen is when a current of 100 milliamps intensity from a 50 to 60 cycle system passes through the cardiac region for a set duration.

Robert: Overhead power lines are often a source of intense shocks, and special precautions are necessary to ensure sufficient clearance, especially for cranes and large excavators. In the event of death from a high voltage accident, it's common knowledge that it occurs through severe burns.

Duncan: Yes, and this is so, because the current passing through the body is often extremely high. However, it's quite remarkable that death doesn't always result from a person being subjected to a very high current. There have been cases where workmen have sustained serious, but not fatal injuries, yet steel angle members in contact with them have literally melted.

Robert: I have known instances of workmen on ladders, or in similar situations, coming into contact with high voltage cables, getting a violent shock, and being thrown to the ground without receiving any more injuries than a few broken bones. In fact, recovery in such circumstances has been quite frequent, and one wonders sometimes how the human frame stands up to this hard treatment.

Duncan: The usual explanation for people getting away with it so lightly in most of the cases you mentioned, is that the likely countershock of the fall often restores the normal working of the heart. Another feasible explanation is that the contact with the cable is generally for only a very limited time.

Robert: By the way, where anyone is rendered unconscious, the "kiss of life" or artificial respiration must be applied immediately. This is one example where first-aid training is vital.

Duncan: I couldn't agree more. Pursuing the subject of electric shock a little further, it's interesting to note that as a result of experiments on animals, it has been proved that a current exceeding 5 amps does not cause fibrillation of the heart. Above this particular amperage apparently, research has found that an immediate contraction of the cardiac muscles occurs, followed by a certain time lapse, after which an impulse from the nervous system controlling the heart causes it to resume its beat.

Robert: Quite an absorbing aspect of this subject. We've already mentioned the danger of overhead cables, but now the question arises as to what a crane operator, for example, should do when the boom of his machine strikes a cable.

Duncan: I must confess, I would be in some doubt myself.

Robert: Well, I might be able to provide a few useful tips here. I hope so, anyway. First of all, one can only assume that he is not in direct contact with the cable, in which case his situation is not as critical as one might imagine. Don't get me wrong, for I'm not suggesting he's sitting pretty, by any means. In most circumstances, the best immediate advice he can have is to stay put and avoid any contact with the cable.

Duncan: Of course, it's easy enough to say that when one is remote from such a nasty situation, but I appreciate that any sensible advice to prevent panic can be helpful.

Robert: If the operator is in an enclosed cabin, however, he should try to drive the crane away from the cable. The best solution, where possible, would be to shout to a colleague to cut off the power.

Duncan: That's fine in theory, Robert, but so seldom can it be done at a moment's notice. It's normally a job for the electricity authority, and that means making a telephone call, which takes time.

Robert: Where the cable has up to 10,000 volts and the crane is wheel-mounted, the rubber tyres will certainly begin to burn.

Duncan: Tell me, if you were an operator faced with this sort of problem, what would you do?

Robert: Well, Duncan' I would jump clear of my machine – that's for certain. Without help, I'd have no option.

Duncan: On the spur of the moment, would you remember the correct way to jump?

Robert: I hope I would. Anyway, that's a good point, and it's worth while knowing how to land on the ground to minimise the danger of the current surrounding the machine. It is important to land with both feet close together, avoiding any ground contact with the hands spread-eagled.

Duncan: For what reason exactly should one land in this particular way?

Robert: To explain why, one has first of all to picture the direction in which the electricity flows from the machine. It would be in the form of a star radiating from the machine, if one were able to see it. At every point from the machine, within a given range, there is a certain electrical potential. The greater the distance between points, the greater the tension between them, and therefore it is safer to come into contact with one point rather than with two some distance apart.

Duncan: Would it be possible to get some help from a colleague, if jumping safely off the machine was out of the question?

Robert: Well, there isn't much that anyone else can do. If a fellow worker felt he must help, which is quite natural, he would have to be very careful how he approached the machine. He should avoid walking or running towards it in the normal way, because he would undoubtly get an electric shock in his legs at least. On no account must he step on to the machine without taking rather special safety precautions.

Duncan: Such as what? He'd have to act pretty smartly to get the operator out, before the machine went up in flames.

Robert: I agree, time is not on his side, but he can't afford to throw his own life away. Once he has reached the machine by hopping rather like a kangaroo, assuming he was sucessful, he would have to be skilful enough to throw some metal object, like a steel ladder or something similar, against the chassis to electrically bridge the machine with the ground. It is then possible to climb safely to the cab.

Duncan: The rescue act doesn't end there, of course, but I think you've given enough information which could be useful to somebody. Less apparent dangers than overhead cables very often occur on sites, and these should not be overlooked.

Robert: What are you driving at?

Duncan: Actually, I'm referring to underground electric and gas mains etc., which can so easily be struck by an excavator bucket during the excavation work.

Robert: This is why it's so important to check the position of burried mains before any work commences. The local authority and public services between them should have plans giving the location of all underground services, and these must be carefully examined.

Duncan: Unfortunately, the records of some of these departments are not always up to scratch, and therefore an element of danger could be lurking around on most sites. In Britain nowadays, it's common practice to lay a plastic warning tape in the trench some distance above a burried high voltage electric cable. This is in addition to the usual clayware cable tiles placed directly over the cable. The idea of the tape is to give the excavator driver some warning that a dangerous cable is lying a little way beneath.

Robert: In closing this discussion, we ought to say in fairness how electricity does contribute to safety, having already highlighted its dangers. Let me therefore expand just a little on the slogan "light the site, speed the work", which I once came across in an English publication. It's true to say that good site lighting helps to cut out risks, simply because a worker can only be safe when he can see where he's going and what he's doing.

Duncan: Incidentally, it also reduces the risk of trespass and theft at night. Remember, no thief likes to be spotlighted.

Robert: Much of our discussion may not have been confined to matters of pure safety, but our readers, I know, will realise that safety crops up in most aspects of construction, perhaps more so with electricity.

Duncan: Yes, that's true. You're happy, I'm sure, with all the electrical topics you've managed to worm into our conversation. I reckon we should have written a chapter on the subject. Don't worry, I'm only pulling your leg.

Robert: To sum up, I don't think any amount of talking on this somewhat unpleasant subject will produce any quick solutions to the problem of preserving life and limb on sites. With the ever progressing mechanisation of construction processes, new dangers rear their ugly head and new safeguards have to be found.

Duncan: This is true, of couse, but one shouldn't overlook the fact that mechanisation has also removed many old dangers, which existed when men were in closer contact with their work. Just imagine the tremendous dangers to which workers of older times were exposed.

Robert: Yes, I've often tried to picture the manual block by block erection of the massive Pyramids, or the large scale hand digging of many of the famous old shipping canals. Our predecessors were probably as aware as we are today of the dangers, and obviously knew something about safety precautions, but I'm sure that the death toll and accident rate was far higher than it is nowadays.

Vocabulary	Wörterverzeichnis	Glossaire
access tunnel	Zugangstunnel	tunnel d'accès
pit	Baugrube	fouille
extension project	Erweiterungsplan	projet d'agrandissement
carpenter	Zimmerer, Zimmermann	charpentier
stretcher	Tragbahre	brancard
to put it bluntly, (to be absolutely honest)	geradheraus sagen	parler carrément
steel beam	Stahlträger	poutre en acier
soldier beam	Rammträger	madrier vertical
hoisting cable	Aufzugseil	câble de levage
"chippy", (nickname for carpenter)	Spitzname für Zimmerer	sobriquet pour charpentier
to carry the can, (to take the blame)	die Schuld auf sich nehmen	en prendre la responsabilité

to be man enough, (to be strong enough)	stark genug sein	être suffisamment fort
to get a move on	schnell machen, beeilen	faire marcher
to run behind schedule	gegenüber dem Baufristenplan im Verzug sein	être en retard par rapport au plan d'avancement
to draw the line, (to stop)	haltmachen	arrêter
slaphappy, (careless)	leichtsinnig	imprudent
flange	Flansch	bride
wear and tear, (results of normal use)	natürliche Abnutzung	usure normale
inspectorate	Aufsichtsbehörde	administration surveillante
site office	Baubüro	bureau de chantier
crane	Kran	grue
lifting equipment	Hebezeug	équipement de levage
excavator	Bagger	excavatrice
tower crane	Turmdrehkran	grue-tour
driving license	Führerschein	permis de conduire
public highway	öffentlicher Verkehrsweg, Landstraße	route
technical control, technical inspectorate	Technischer Überwachungsverein, T.Ü.V.	service public de contrôle des matériels
hired plant	fremdgemietete Baumaschinen und -geräte	machines et engins loués
workshop facilities	Werkstatteinrichtung	installation d'atelier
trained mechanic	Maschinenschlosser	mécanicien
maintenance	Wartung, Unterhaltung	entretien
repair	Reparatur	réparation
to carry, (to keep or to have)	"dahaben"	avoir
set-up, (organisation)	Organisation	organisation
to run a machine into the ground, (to work a machine really hard until it can work no more)	eine Maschine bis zur Unbrauchbarkeit (über)beanspruchen	surcharger une machine jusqu'à sa défaillance totale
reconditioned machine	überholte Maschine	machine reconditionnée
first-aid training	"Erste-Hilfe"-Kursus	cours de premier secours
walk of life, (job or profession)	Beruf	profession
site lighting	Baustellenbeleuchtung	éclairage de chantier
power	Kraft, Energie	force, énergie
electrically minded, (having electrical knowledge)	über Elektrizität Bescheid wissen	connaissances en électricité
wiring up, (fixing or installing cables etc.)	elektrische Leitungen installieren	installation des conduites électriques
insulation	Isolation	isolation
plug	Stecker	fiche de prise de courant
junction box	Verteilerkasten, Stromverteiler	boîte de distribution
positioning of cables	Anordnung von elektrischen Leitungen	positionnement de câbles électriques
working plant	Arbeitsmaschinen	matériels
tall order, (difficult or impossible request)	"harte Nuß", schwer erfüllbare Forderung	ordre difficile à exécuter
heck of a long way	"verflixt" weiter Weg	chemin diablement long
C.E.E.(International Commission on Rules for the Approval of Electrical Equipment)	Internationale Kommission für Regeln zur Begutachtung Elektrischer Erzeugnisse	Commission Internationale de réglementation en vue de l'Approbation de l'Equipement Électrique

point, pin	Steckkontakt, Steckerstift	broche de contact
non-interchangeable	(pol)unverwechselbar, nicht austauschbar	non interchangeable
to standardise	vereinheitlichen, normen	normaliser
sweeping change, (large scale or extensive change)	umfassende Änderung	changement profond
socket outlet	Steckdose	prise de courant
to chip in, (to interrupt)	ins Wort fallen, einwerfen	interrompre
earth terminal pin	geerdeter Polstift	broche de mise à la terre
live pin	stromführender Steckerstift	broche sous tension
to engage	kuppeln, einstecken	engager
shutter	Verschluß(vorrichtung)	fermeture
fuse	Sicherung	fusible
to blow a fuse	eine Sicherung durchbrennen	faire sauter un fusible
circuit	(Strom-)Kreis	circuit
switch	Schalter	interrupteur
"pros and cons"	"Pro und Kontra"	pour et contre
scraping earth contact	Erd-Schleifkontakt	terre par contacts latéraux
voltage	Spannung	voltage, tension
to outline, (to give the basic idea)	umreißen	montrer les grandes lignes
to take something up, (to talk about something)	sich einer Sache annehmen	se charger de quelquechose
network	(gesamtes) Versorgungsnetz	réseau de distribution
power supply	Energieversorgung	alimentation en force
Electricity Board	Elektrizitätsgesellschaft, Elektrizitätsbehörde	compagnie d'électricité
frequency cycle	Frequenz, Periode	fréquence
concrete poker	Betoninnenrüttler, Rüttelflasche	pervibrateur
electric tool	Werkzeug mit elektrischem Antrieb	outil à entraînement électrique
insulated casing	Isoliergehäuse	boîtier isolant
transformer	Transformator, Wandler	transformateur
power tool	Krafthandwerkzeug	outils mécanique
steel rod	Stahlstab	barre d'acier
earth clamp	Erdungsklemme	pince de mise à la terre
earth-leakage circuit breaker	Fehlerstrom-Schutzschalter, FI-Automat	interrupteur de fuite disjoncteur différentiel
alternating current, (a.c.)	Wechselstrom	courant alternatif
electric shock	elektrischer Schlag	choc électrique
resistance	Widerstand	résistance
fibrillation of the heart	Herz(klappen)flimmern	palpitation des valvules du coeur
overhead power line	(Hochspannungs-) Freileitung	ligne aérienne haute tension
clearance	Freiraum(höhe)	hauteur de libre passage
steel angle member	Winkelstahl-Bauteil	élément cornière en acier
human frame, (body)	Körper	squelette humain
kiss of life	Mund-zu-Mund-Beatmung	respiration artificielle de bouche à bouche
artifical respiration	künstliche Beatmung	respiration artificielle
amperage	Stromstärke	ampérage
crane operator	Kranführer	grutier
boom	Ausleger	flèche
to get someone wrong, (to misunderstand someone)	jemanden mißverstehen	mal comprendre quelqu'un
to sit pretty, (to be free from trouble)	"aus dem Schneider sein", außer Gefahr sein	être sorti d'affaire

to stay put, (not to move at all)	an Ort und Stelle bleiben	rester sur place
enclosed cabin	geschlossenes Führerhaus	cabine
electricity authority	Elektrizitätsbehörde	administration de l'électricité
wheel-mounted	auf Rädern	sur roues
rubber tyre	Gummireifen	pneu
on the spur of the moment, (without having time to think)	ohne Überlegung, unter dem ersten Eindruck	instinctivement
spread-eagled, (spread apart)	ausgebreitet, gespreizt	étalé, écarté
tension	Spannung	tension
fellow worker	Arbeitskamerad, Arbeitskollege	collègue
cab	Führerstand	poste de commande
underground main, (underground service)	unterirdische Haupt(versorgungs)leitung	conduite principale souterraine
excavator bucket	Baggerschaufel, Baggerlöffel	pelle d'excavatrice
excavation work	Aushubarbeit	travaux d'excavation
local authority	Ortsbehörde	autorité locale
public services, (public utilties)	öffentliche Versorgungsbetriebe	services publics
record	Bestandsaufnahme, Aufzeichnung	liste, relevé
up to scratch, (in first class order)	auf der Höhe	de premier ordre
tape	Band	bande
trench	Graben	tranchée
clayware cable tile	Kabelabdeckziegel	tuile de couverture de câble
to highlight, (to emphasize)	ausdrücklich hervorheben, betonen	souligner
trespass	unbefugtes Betreten	entrée interdite
to spotlight	mit Scheinwerfer anleuchten	projecteur
to worm in, (to slip in bit by bit)	sich "hineinmogeln"	se glisser petit à petit
to pull someone's leg, (to be joking)	Spaß machen, scherzen	être de bonne humeur
to preserve life and limb	Leib und Leben schützen	préserver la vie
safeguard	Sicherheitsmaßnahme	mesures de sécurité
large scale hand digging	Erdarbeiten großen Maßstabes von Hand	travaux de terrassement manuels de grande ampleur
death toll	Blutzoll, Todesrate	perte de vie humaine
accident rate	Unfallrate	pourcentage d'accidents

11. DAMS

In the broadest possible sense, dams fall into two main categories, namely those which prevent the sea, rivers or canals from flooding the adjoining land, and those which form impounding reservoirs in which water is stored for supply purposes, and often for generating electricity as well. When the intention is chiefly to hold back flood water, dams are usually called dykes or levées, which are, in fact, simple embankments. The actual term dam is nearly always associated with the larger structures for storing water, although the structural principle is basically the same for all means of retaining water. Dams in the true sense constitute two forms constructionally, one being shaped either rather like a triangle or trapezium and built up of earth, loam and rock filling, whilst the other is far more slender and normally constructed of concrete.

The principal engineering problems, which have to be overcome before the construction of a dam begins, are providing enough resistance against overturning and sliding under the force of water, and ensuring that the dam is sufficiently watertight. These problems are more acute with modern dams, which are generally constructed with greater effective heights, and therefore have to withstand tremendous water pressure. In many projects too, the diversion of the existing river running through the valley often requires much thought. Where the valley is really a narrow canyon, tunnels may be necessary as the only effective means. Whatever method of diversion is adopted, it must be carried out well in advance of the main project, so as to avoid any disruption of the site organisation.

Earth dams and so-called gravity dams must be made very heavy with their centre of gravity as low as possible. A high degree of resistance against sliding is most important with these particular types of dam, and to acheive this, care must be taken to ensure that the base of the dam, as well as the foundation, is extremely rough. It is not usually a feasible proposition to construct a high earth or gravity dam to retain a massive volume of water, since it would have to be exceedingly heavy and would cover a vast area. Both of these factors would involve unnecessary expence.

Because of their comparative simplicity, earth dams are often a practical alternative to the more sophisticated concrete dams in situations where low dams are more suitable. The materials used in the construction of an earth dam are comparatively cheaper and easier to transport, even though they are in huge quantities. They are usually sand, gravel, rock and clay. Fig. 22, illustrates an earth dam made up of typical layers of various materials. The main part is of compacted rock filling, which provides the necessary weight and stability. The core is of clay, and this together with a layer of impervious material next to the retained water should make the dam sufficiently watertight. To protect this layer, gravel or rock chippings are usually laid against it, and these are then covered with a layer of concrete or asphalt. The remaining materials and construction method shown in this Fig. are really added measures to give further stability.

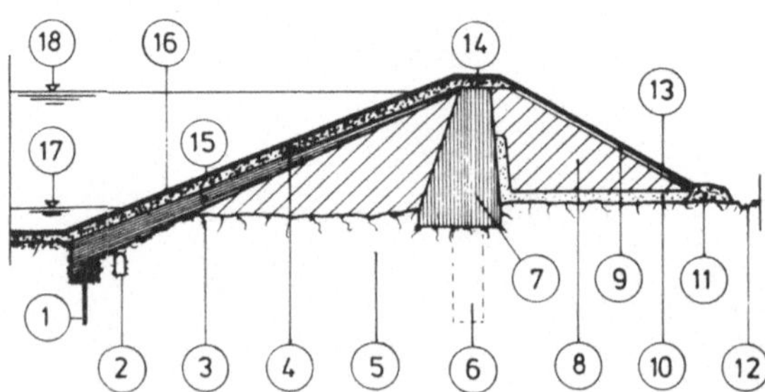

Fig. 22: Earth Storage Dam

earth storage dam	Erdstaudamm	barrage de retenue en terre, barrage en terre

Where a dam is to be constructed across a deep narrow canyon, the arched type is normally the most suitable. The wall, forming the greater part of the dam, is convex on plan on the side in contact with the water, and this design transmits most of the pressure to the rock sides of the canyon. It is often a considerable advantage to have this wall curved in section as well, thus forming a shell-like structure, which helps to save material and provide a better grip at the base. To avoid failure in an arched dam, which is largely supported by the thrust from the sides of the canyon serving as abutments, it is essential to construct it on solid rock and never on yielding ground. If the rock is not solid enough, all gaps and fissures must be filled either with a cement grout, a bituminous injection or similar material to improve its stability and density. Although an arched dam is far more elegant in appearance than the rather clumsy gravity dam, a form of construction known as a gravity-arch dam, which embodies features of both types, is often adopted, where suitable, because of its greater stability, and in cases where the supporting conditions are suspect.

A multi-arch dam can be constructed by having a series of convex walls spanning between buttresses, which are in effect relatively short lengths of thick walling. The buttresses afford additional strength to resist pressure which is conveyed to them in the form of arch thrust. Since this type of dam is comparatively lightweight, it is particularly appropriate where the foundation has a poor bearing capacity. Where flat reinforced concrete slabs are used instead of arched concrete walling between buttresses to form the face against the water, such dams are known as flat-slab deck dams or flat-slab buttress dams. These are usually laid at an angle of about 45 degrees.

All dams which are constructed primarily to store water have two features in common, namely a sluiceway through the bottom part and a spillway near the crest of the dam. The sluiceway must be situated at the lowest point to allow the reservoir to be completely drained for cleaning or repair work. Both means of

1	**sheet piling**	Spundwand	rideau en palplanches
2	**sewer**	Entwässerungskanal	canal d'assèchement
3	**layers of impervious material**	Lagen undurchlässigen Materials	couches imperméables
4	**coarse gravel, ballast or chippings**	Grobkies, Schotter oder Splitt	gravier grossier, pierre concassée gros calibre ou gravillon
5	**bedrock**	anstehender Fels, gewachsenes Gestein	roche saine
6	**grout curtain**	Mörtelschleier, Verpreßmörtelzone	écran d'étanchéité
7	**impervious core of clay or mixture of sand and silt**	undurchlässiger Kern aus Ton (Lehm) oder einem Feinsand-Schluffgemisch	noyau imperméable en argile ou avec mélange de sable fin
8	**compacted rock fill**	verdichtete Steinschüttung	enrochement compacté
9	**semi-pervious blinding coat**	halbdurchlässige Randzonenlage	couche de propreté semi-perméable
10	**graded filter layer**	mehrstufige Filterschicht	couche filtrante graduée
11	**riprap**	Steinschüttung (am Böschungsfuß)	enrochement
12	**drainage channel, drainage ditch**	Entwässerungsgraben	canal d'assèchement
13	**top soil with turfing**	Mutterboden mit Rasendecke	sol végétal avec couverture de gazon
14	**crest level, top of the dam**	Kronenhöhe	hauteur de crête
15	**upstream slope**	oberwasserseitige Böschung	talus côté amont
16	**asphaltic concrete or stone paving slabs**	Asphaltbeton oder Steinplattenbelag	béton asphaltique ou revêtement en dalles de pierre
17	**lowest water level (l.w. l.)**	niedrigster Wasserstand	niveau d'eau minimum
18	**highest water level (h.w. l.)**	höchster Wasserstand	niveau d'eau maximum

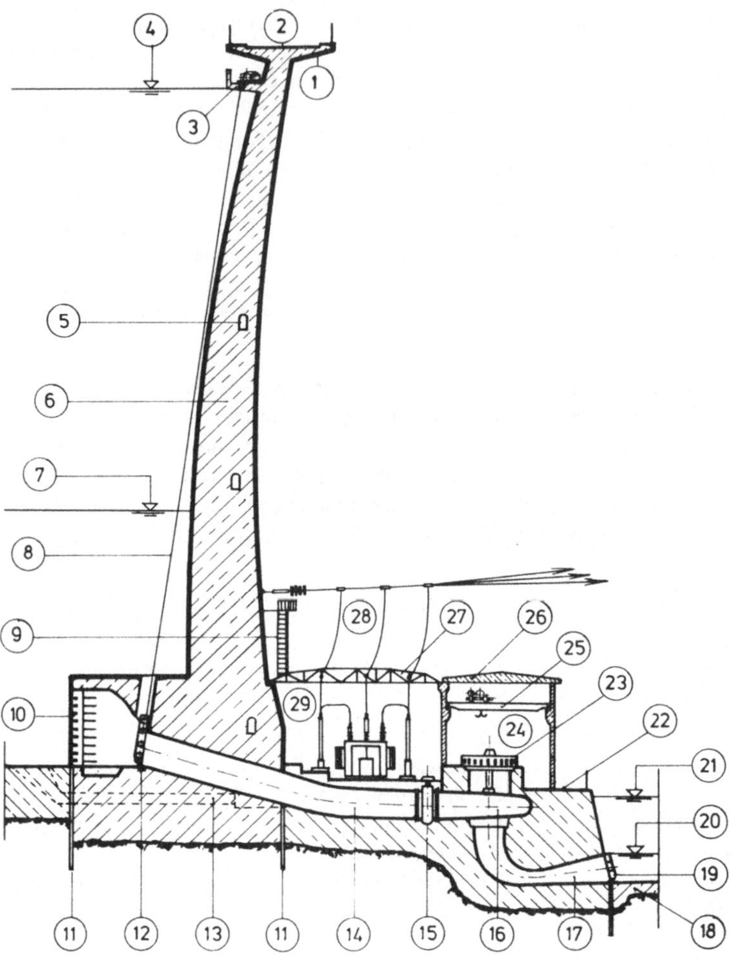

Fig. 23: Arch Dam with Power House and Transformer Station

	arch dam	Bogenstaumauer	barrage-voûte
	power house	Maschinenhaus	bâtiment des machines
	transformer station	Umspannanlage	station de transformation
1	**cantilevered roadway with footpaths**	ausgekragte Fahrbahnplatte mit Gehsteigen	dalle de chausée en encorbellement avec trottoirs
2	**dam crest**	Talsperrenkrone	crête de barrage
3	**control platform with winch**	Bedienungsbühne mit Seilwinde	plate-forme de commande avec treuil
4	**highest water storage level**	Stauziel	retenue max.
5	**inspection gallery**	Beobachtungsstollen	galerie de contrôle
6	**arched concrete dam**	Bogenstaumauer	barrage du type poids-voûte
7	**lowest operating water level**	Absenkziel	vidange minima
8	**intake gate control cable**	Bedienungsseil für Einlaßverschluß	câble de commande de vanne d'entrée
9	**observation platform access ladder**	Stiege zur Beobachtungsplattform	escalier d'accès à la plate-forme d'observation
10	**intake screen**	Einlaufrechen	grille d'entrée
11	**sheet piling**	Spundwand	cloison en palplanches

12	**intake lifting gate**	Einlaß-Hubschütz	vanne d'entrée à mouvement vertical
13	**low level sluiceway**	Grundablaß	évacuateur de fond
14	**penstock**	Druckleitung, Triebwasserleitung	conduite forcée
15	**throttle valve**	Drosselventil	soupape d'étranglement
16	**turbine housing**	Turbinengehäuse	carcasse de turbine
17	**tapered tailrace**	Auslauftrompete	cône de fuite
18	**base to adjacent stilling basin**	Sohle des anschließenden Tosbeckens	radier du bassin de tranquillisation adjacent
19	**non-return flap valve**	Rückstau(verschluß)klappe	vanne de non-retour
20	**normal tailwater level**	normaler Unterwasserstand	niveau normal d'eau aval
21	**highest tailwater level**	höchster Unterwasserstand	niveau maxi d'eau aval
22	**access road**	Zufahrtstraße	voie d'accès
23	**generator**	Stromerzeuger	génératrice
24	**power house**	Krafthaus, Maschinengebäude	bâtiment des machines
25	**overhead travelling crane with crab**	Laufkran mit Laufkatze	pont roulant avec chariot de roulement
26	**solid reinforced concrete roof slab**	massives Stahlbetondach	toit en béton armé
27	**steel roof truss construction**	Stahlfachwerk-Dachkonstruktion	fermes en acier
28	**overhead high tension transmission line**	Hochspannungs-Freileitung	ligne aérienne haute tension
29	**transformer and switchgear plant**	Umspann- und Schaltanlage	poste extérieur

discharging water are normally regulated by gates, the sluiceway, for example, being closed with either a roller gate, a slide gate, a shut-off rotary valve, a follower-ring valve or a radial gate. Because of the terrific force of the discharging water, a means of dispersing the energy produced must be installed immediately behind the outlet. This is often achieved by providing stilling pools with countersills, dragon teeth or baffle blocks acting as dissipators.

To prevent flooding from an impounding reservoir, automatic spillways are essential. Besides the ordinary non-adjustable spillways, others have crest gates incorporated, by which the water level is controlled. As with weirs, the type of gate differs from one construction to another, there being, for example, radial, roller, sector, sliding and tilting gates. For less important purposes, or as an additional means of controlling the water level, stop logs are also used. They simply consist of a number of horizontal timber planks, steel joists or reinforced concrete beams placed one on top of the other and held in position on both sides in vertical grooves formed in the structure.

The power installation of a hydroelectric dam can easily become damaged through sudden pressure shocks, and this could occur, for example, when the throttle valve in front of the turbine must be closed quickly. To avoid any such damage, surge shafts, surge chambers or surge tanks are usually connected to the pressure pipes conveying water to the turbines. The relatively simple surge tanks of the older dams have largely given way to more elaborate systems, such as that provided by Johnson's differential surge tank, which reduces the kinetic energy of the water in a shorter time.

Although a cofferdam is nearly always a temporary structure to exclude water from a site to permit reasonably dry working conditions, it does come within the category of dams. Some cofferdams are just earth dams made up of a single material, but others are formed with two sheet piling walls, usually connected by anchor rods or cross walls and filled in between with sand or gravel. A more elaborate and efficient type, known as a cellular cofferdam, consists of flat sheet piles arranged to form circles, which are linked together in a row, rather like a chain, to give increased stability. This form of cofferdam can even be constructed freestanding on a hard bed, into which it is not possible to drive piles. Cofferdams are commonly employed in a variety of situations, and they are particularly useful, for example, in connection with the construction of abutments and supporting piers for bridges over rivers or for tunnels through waterways. A classic illustration of the use of two large cofferdams was across the valley of the river Zambesi at Cabora Bassa in Mozambique, where one of the most extensive hydroelectric schemes was sited. One cofferdam was constructed upstream and the other downstream, so cutting off the river, which was diverted through by-pass canals.

Of all the many hugh civil engineering projects, the construction of a dam must rank amongst the most important and satisfying. In the Far East with its prolonged periods of drought and short periods of high floods, the need for extensive water storage facilities to collect the limited heavy rainfall from the cathment areas is still extremely urgent. The proper collection of rainwater to prevent flooding and the subsequent use of the stored water for irrigation schemes etc. can be largely achieved with dams.

Vocabulary	Wörterverzeichnis	Glossaire
dam	Damm, Sperrmauer, Talsperre	barrage
impounding reservoir, storage reservoir	Speicherbecken	réservoir d'accumulation
water supply	Wasserversorgung	alimentation en eau
to generate electricity	elektrischen Strom erzeugen	produire du courant électrique
flood	Flut, Überschwemmung, Hochwasser	crue
dyke	Deich	digue
levée, embankment	Uferdamm	levée
trapezium	Trapez	trapèze
loam	Lehm	glaise
rock filling	Steinschüttung	enrochement
overturning	Kippen	basculage
sliding	Gleiten	glissage
watertight	wasserdicht	étanche, imperméable à l'eau
diversion	Umleitung	déviation
canyon	tiefe Bergschlucht	gorge
site	Baustelle	chantier
earth dam, earthen dam	Erdstaudamm	barrage en terre
gravity dam	Schwergewichtsmauer	barrage-poids
centre of gravity	Schwerpunkt	centre de gravité
base of the dam	Dammfuß	base du barrage
foundation	Gründung, Fundament	fondation
gravel	Kies	gravier
rock	Fels, Stein	roche
clay	Ton	argile
impervious	undurchlässig	imperméable
chippings	Splitt	gravillon
arched dam	Bogen(stau)mauer	barrage-voûte
on plan	im Grundriß	dans le plan horizontal
friction	Reibung	friction, frottement
grip	Halt, Haftung, Kraftschluß	adhérence
thrust	Schub	poussée
abutment	Widerlager, Kämpfer	culée
yielding ground	nachgiebiger Boden	sol élastique
gap	Spalt	fente
fissure	Riß	fissure
cement grout	Zementmörtel	mortier de ciment

gravity-arch dam	Bogengewichtsmauer	barrage poids-voûte
multi-arch dam	Gewölbereihen-Staumauer Vielfachbogen-Mauer	barrage à voûtes multiples
buttress	Strebepfeiler	pilier d'arc-boutant, contrefort
arch thrust	Gewölbeschub	poussée de la voûte
bearing capacity	Tragfähigkeit	force portante
flat concrete slab	ebene Betonplatte	dalle horizontale en béton
arched concrete walling	gebogene Betonwand	barrage-voûte en béton
flat-slab deck dam, flat-slab buttress dam	Platten(pfeiler)staumauer	barrage à contreforts à dalle plane
sluiceway	Grundablaß	évacuateur de fond
spillway	Überfall	déversoir
(dam)crest	(Damm-)Krone	couronnement, crête
to drain	entwässern, entleeren	drainer
to discharge	ablassen, ausfließen lassen	vider
gate	Verschluß, Schütz	vanne
roller gate	Rollschütz	vanne cylindrique
slide gate, sliding gate	Gleitschütz	vanne à glissières
valve	Schieber, Ventil	vanne
shut-off rotary valve	Kugelabschluß-Schieber	vanne à obturateur sphérique
follower-ring valve	Keilschieber	vanne à lunette
radial gate, sector gate	Segment-Schütz Sektor-Schütz	vanne à segment,
stilling pool	Tosbecken	bassin de tranquillisation
countersill	Gegenschwelle	contre-seuil
dragon tooth	Drachenzahn	dent de tranquillisation
baffle block	Dämpfer-, Ablenk-, Prallblock	bloc chicane
dissipator	Verteiler (der Energie)	dissipateur **(d'énergie)**
crest gate, spillway gate	Überlaufverschluß	vanne de déversoir
weir	Wehr	barrage
tilting gate, automatic flap gate	(Gleichgewichts-) Klappschütz	clapet équilibré
stop log, flashboard	Dammbalken	poutrelle
plank	Bohle	**madrier**
steel joist	Stahlträger	poutre en acier
reinforced concrete beam	Stahlbetonbalken	poutre en béton armé
groove	Nut	rainure
hydroelectric dam	Wasserkraft-Talsperre	barrage de centrale électrique
pressure shock	Druckstoß, Schlag	choc
throttle valve	Drosselventil	soupape d'étranglement
surge shaft	Schwallschacht, Schacht-Wasserschloß	puits de la cheminée d'équilibre
surge chamber	Schwallraum	chambre d'équilibre
surge tank	Wasserschloß	chambre d'équilibre
pressure pipe, penstock	Druckrohr, Druckleitung	conduite forcée
differential surge tank	Differential-Wasserschloß	cheminée d'équilibre différentielle
cofferdam	Fangedamm	batardeau
sheet piling wall	Spundwand	cloison de palplanches
cellular cofferdam	Zellenfangedamm	batardeau à cellules
sheet pile	Spundbohle	palplanche
hard bed	harter Untergrund	terrain dur
supporting pier	Tragepfeiler	pile

upstream, downstream	stromauf, stromab	amont, aval
by-pass canal	Umgehungskanal, Umlaufstollen	canal de deviation
drought	Dürre, Trockenheit	sécheresse
water storage facilities	Einrichtungen zur Wasserspeicherung	installation d'accumulation d'eau
catchment area	Einzugsgebiet	bassin versant
irrigation scheme	Bewässerungsplan	plan d'irrigation

Dialogues

Robert: Tell me, Duncan, do you share my view that dams have an ever increasing importance in the field of civil engineering?

Duncan: Well, if you mean those mainly intended for the production of energy, I can't say I do entirely. You see, Robert, on the whole I feel that the development of nuclear power stations has more far-reaching possibilities.

Robert: For what reason, exactly?

Duncan: For a start, it is usually possible to site a nuclear power station reasonably close to industrial and densely populated areas, where most energy is required. The majority of hydroelectric dams, on the other hand, have to be constructed in the wilds with all the problems of access and difficult working conditions. It's no joke carrying out an extensive project in the mountains, for example, and so I think that fewer dams are likely to be planned in the future principally for the generation of electricity. I could be mistaken, of course.

Robert: Good heavens! I didn't realise that our opinions differed so widely on this subject. I can see now, that the little chat that I had envisuaged might well develop into a somewhat deeper discussion. Maybe we should firstly try to clear up a fundamental misunderstanding of the very term 'dam', which even occurs between old friends like you and I. However, I hardly think this is the root of our different points of view.

Duncan: Ah yes, there's quite a difference in the meaning of the English 'dam' and the German 'Damm', even though the words are very similar. Your 'Damm' refers only to an earth dam, I believe, whereas our 'dam' covers all forms of dam, including the massive hydroelectric dam.

Robert: That's perfectly true. When I expressed the view earlier, that dams are growing in importance, I was certainly thinking more particularly of those in connection with hugh water regulation schemes, like the well known Delta Plan in the Netherlands.

Duncan: In such projects, I would totally agree that dams have unlimited scope, since prevention against flooding, often causing the salination of good agricultural land, and reclamation of coastal stretches are both high on the list of priorities. Now, turning specifically to earth dams, are there likely to be any real prospects for them in the years to come? They haven't changed much, to my mind, since Roman times, and I have mixed feelings about their future, even though many recent dams of note have been rock-filled earth dams.

Robert: History isn't one of my strongest subjects, as you well know, but I'm convinced that earth dams have improved quite a lot since those far off days. When you say there's been little change, you have obviously only considered the general appearance and basic construction principle. In my opinion, the modern form still has much to offer.

Duncan: Ah, that may be so, but aren't you evading my question about its real prospects?

Robert: Give me chance, I'm just coming to that. Today's detailed soil investigation in relation to bearing capacity, density, penetration, friction and cohesion goes a long way to ensuring that an earth dam is more reliable and economical to construct than it was many years ago.

Duncan: Well, I can't argue with you on that fact. To be truthful, it isn't only modern soil investigation that has given the earth dam a bit of a boost, but the improved methods of grading, compacting and stabilising the filling materials. Even so, I don't think the risk of core cracks has been entirely eliminated yet.

Robert: You seem to have changed your tune a little about the possibilities of earth dams, or have you merely had second thoughts about their merits?

Duncan: At the moment, no doubt, there are still valid reasons for their continued use in certain circumstances, but in the long term, I believe they will vanish from the scene.

Robert: In what circumstances, as a matter of interest, would you consider that they are justified?

Duncan: Where a good supply of suitable filling material is readily available from a local quarry, or the spoil from the excavation work on the project could be utilised as a ready-made filling, then I would say it was logical to consider and probably construct an earth dam, provided, of course, that all other governing factors are favourable.

Robert: Forgive me, Duncan, if I wander a little from this topic, but it's well worth remarking that a great deal of knowledge has been gained regarding soil behaviour through construction failures. Many theories, in fact, on soil statics are based on the lessons learnt from broken dams. It may seem incredible, but I can recall a case where a consulting engineer even became quite famous as an indirect result of the failure of a dam which he had designed. The loss of life and damage caused was considerable.

Duncan: In the circumstances, how in heaven's name did he manage to improve his image?

Robert: Well, it wasn't because of his mistake, but through his ability afterwards to produce a more reliable theory relating to the shear failure of soil in an embankment.

Duncan: Life certainly has its peculiar twists and turns and one can never tell what good may emerge from the most terrible disasters. To come back to earth dams, I believe that today's highly developed earth-moving plant has certainly influenced their continued us. One need only to be reminded of suction dredgers, which are often used to form harbour basins and at the same time build up dams or embankments with the dredged material to retain the water.

Robert: I've been turning over in my mind what you said earlier on about nuclear power stations being generally more feasible than hydroelectric dams. Whilst I accept this up to a point, I can't see in the foreseeable future water power not being harnessed to the full. In my experience, most concrete wall-dams are constructed mainly, if not solely to form part of a reservoir for fresh water to supply our cities and industrial areas. At the same time, such dams are often used to regulate the water level of waterways, which in consequence controls the level in adjoining agricultural land. With the collection and storage of much water, it's only common sense to turn it into energy, where practical. This brings us to an interesting way of using stored water for energy. I'm thinking of the system employed in pumped storage plants which are, incidentally, growing in importance.

Duncan: Yes, I know what you mean. Two reservoirs are constructed close together, one at a higher level than the other, and during the off-peak night period electric energy is used to pump the water from the lower to the higher reservoir. In the day-time during the peak period of electric consumption, the flow of water is reversed, that is from the higher to the lower reservoir, thus generating electricity

Robert: You're dead right. What happens, in fact, is that the night-time pumping station becomes the generating station in the day-time. A very economical process, I would say.

Duncan: At the moment, there are many vast schemes all over the world, either in construction or in the planning stage, to store water when it's available, such as during the rainy season or in the spring-time when the snow is melting in the mountains. By so doing, water is at hand in the dry season to irrigate the land, provide the population with a water supply, and keep the rivers and canals flowing for water traffic.

Robert: I'm prepared to stick my neck out and forecast immense world-wide programmes to control water and direct it from areas where it is in abundance to other areas starved of it. Just look at the huge areas of desert on our globe which are crying out for water, so that better use can be made of them. Then look at those other regions flooded with water and often useless as a result. The battle to put water to better use has, in my view, only just begun. What a challenge for our civil engineers!

Duncan: I'm sure you are right, and this is where dams play a major part in such programmes. Electric energy could well become a by-product of many future large schemes, rather then the chief reason for having them, as has sometimes been the case in the past.

Robert: Don't forget too, that water consumption in the world, especially in our towns and for industry is increasing dramatically, and this in itself warrant many more reservoirs.

Duncan: Maybe we haven't always seen eye to eye on the subject of dams, but in respect of your last remarks, I'll back you up to the hilt.

Vocabulary	Wörterverzeichnis	Glossaire
civil engineering	Ingenieurbau	génie civil
nuclear power station	(Atom)-Kernkraftwerk	centrale électrique nucléaire
industrial area	Industriegebiet	zone industrielle
densely populated area	dicht besiedeltes Wohngebiet	zone d'habitation à population dense
in the wilds	in der freien Natur	en pleine nature
to be no joke	nicht angenehm sein	ne pas être amusant
root	Grund, Ursache	origine
water regulation scheme	Wasserbe- und Entwässerungsprojekt	project de régularisation des eaux
salination	Versalzung	ensalement
agricultural land	landwirtschaftlicher Boden	terre arable
reclamation	(Wieder-)Gewinnung, Nutzbarmachung	défrichage
coastal stretch	Küstenstrich	littoral
soil investigation	Bodenuntersuchung	investigation du sol
density	Raumgewicht	densité
penetration	Eindringung	pénétration
cohesion	Kohäsion	cohésion
bit of a boost	ein wenig Auftrieb	légère poussée verticale
grading	Abstufung nach Korngröße	graduation
compacting	Verdichten	compactage
stabilising	Festigen	stabilisier
filling material	Füllmaterial	matériau de remplissage
core crack	Kernriß, Kernbruch	fission, cracking
to have second thoughts	näher nachdenken	y réfléchir à deux fois
quarry	Steinbruch	carrière
spoil, excavated material	Aushub	matériau excavé
ready-made	(gebrauchs)fertig	prêt à l'usage
soil behaviour	Bodenverhalten	comportement du sol
soil statics	Bodenmechanik, Erdbaustatik	mécanique des sols
consulting engineer	beratender Ingenieur (Ingenieurbüro)	ingénieur-conseil
shear failure of soil	Grundbruch	renard
earth-moving plant	Erdbewegungsmaschinen und -geräte	engins de terrassement
suction dredger	Saugbagger	drague suceuse
harbour basin	Hafenbecken	bassin portuaire
to turn something over in one's mind (to think about something)	wiederholt durch den Kopf gehen lassen	bien y réfléchir
to harness	nutzbar machen	utiliser
pumped storage plant	Pumpspeicherwerk	usine de pompage
off-peak period	belastungsspitzenfreie Zeitspanne	période sans charge de pointe
peak period	Spitzenbelastungszeit	période de pointe
generating station	Kraftwerk, Eltwerk	centrale électrique
planning stage	Entwurfsstadium, Planungsphase	stade des études
to irrigate	bewässern	irriguer
water traffic	Schiffsverkehr	trafic de bateaux
to stick one's neck out, to take a chance	es darauf ankommen lassen, riskieren	risquer

to cry out for something (to demand something)	dringend nach etwas verlangen	exiger quelquechose
by-product	Nebenprodukt	sous-produit
to see eye to eye (to agree)	übereinstimmen, gleicher Meinung sein	être du même avis
to back someone up to the hilt (to fully support someone)	jemanden voll unterstützen	soutenir quelqu'un sans restriction

(Reproduced by permission. Master Agency U.S.A.)

12. ABBREVATION OF CONSTRUCTION TERMS

In Britain, abbreviations are mainly used by surveyors when taking off quantities from drawings for the preparation of bills of quantities, and when recording site measurements for the preparation of a final account. The actual bills of quantities and final account themselves are always written in full. Many of the more common abbreviations, e.g. d.p.c. (damp proof course) or g/f (ground floor), are sometimes used by architects and engineers in marking up their drawings.

The time-saving abbreviation of technical language, which is second nature to the British quantity surveyor, may well cause some difficulty in understanding at first, and therefore a little explanation is necessary. Many words, particularly verbs, are often left out when the meaning of the shortened sentence is obvious without them. However, no word should ever be omitted if its omission might cause any ambiguity.

A good example of the legitimate omission of a verb ("build" in this case) is given in the first item of Section D – Brickwork and Blockwork. Here, there is clearly no need to say "build 225 mm brickwork", since it is obvious that the brickwork cannot be produced without building it. On the other hand, the fourth item of the same section is an example where the verb "build in" cannot possibly be omitted, because the item mainly describes only the action of building in the end of a bearer to blockwork.

EXAMPLES OF COMMON BRITISH ABBREVIATIONS

Trade	Abbreviation	Full Meaning
A		
Demolitions	B.u. conc fdns & rem.	Break up concrete foundations and remove.
	B.o. 150 mm c.i. dr & rem.	Break out 150 mm cast iron drain and remove.
	T.u. st pavg, cl & s.a.	Take up stone paving, clean and set aside.
	P.d. 225 mm brwk & rem.	Pull down 225 mm brickwork and remove.
	C.o. 450 mm X 225 mm vent in 275 mm h.w. & m.g. opg wi ct. rend.	Cut out 450 mm X 225 mm ventilator in 275 mm hollow wall and make good opening with cement rendering.
	T.o. r.l. & furn & r.t.s.	Take off rim lock and furniture and return to store.
	H.o. plas on blwk & rem.	Hack off plaster on blockwork and remove.
	T.d. ptn p/bdd b/s & rem.	Take down partition plasterboarded both sides and remove.
B		
Excavation	Exc o.s. av 150 mm dp & dep in sp hps n.e. **100 m dist.**	Excavate over site average 150 mm deep and deposit in spoil heaps not exceeding **100** m distance
	Exc to r/l.	Excavate to reduce levels.
	Exc tr for fdns.	Excavate trench for foundations.
	E.o. for exc in hd mat.	Extra over for excavating in hard material.
	L. & rem exc mat o.s.	Load and remove excavated material off site.
	R.f. & r. ard fdns.	Return, fill and ram excavated material around foundations.

L. & r. btm exc to rec h/c	Level and ram bottom of excavations to receive hardcore.
Exc in sp hps rem a dist n.e. 50 m & dep, s. & l. to m.u. lvls ard bldgs.	Excavate in spoil heaps, remove a distance not exceeding 50 m and deposit, spread and level to make up levels around buildings.
P. & s. sides of bsmt exc n.e. 3 m dp.	Plank and strut sides of basement excavation not exceeding 3 m deep.
100 mm h/c bed s.l.r. & c. incl blg wi fine mat to rec o/s conc.	100 mm hardcore bed spread, levelled, rolled and consolitated including blinding with fine material to receive oversite concrete.

C
Concrete Work

R.c. (1 : 2 : 4) grd bm exc 0.05 m^2 & n.e. 0.10 m^2 s.a.	Reinforced concrete (1 : 2 : 4) ground beam exceeding 0.05 m^2 and not exceeding 0.10 m^2 sectional area.
150 mm pl conc (1 : 2 : 4) g/f sl fin to rec c. & s. bed.	150 mm plain concrete (1 : 2 : 4) ground floor slab finished to receive cement and sand bed.
Trow conc sm as pavg.	Trowel concrete smooth as paving.
Prec conc (1 : 2 : 4) lin 225 mm × 150 mm fin f. on exp surfs & reinfcd wi & incl 2 No. 12 mm m.s. reinfcg rods.	Precast concrete (1 : 2 : 4) lintol 225 mm × 150 mm finished fair on exposed surfaces and reinforced with and including 2 No. 12 mm mild steel reinforcing rods.
Wrot f/wk to s. & s. conc bm.	Wrought formwork to sides and soffites of concrete beam.
E.o. f/wk for fillet p.o. to form sm thrtg in u/s conc wdw hd.	Extra over formwork for fillet planted on to form small throating in underside of concrete window head.
Mort in conc landg for end of 25 mm × 25 mm bal & g.i. wi c.m.	Mortice in concrete landing for end of 25 mm × 25 mm baluster and grout in with cement mortar.
Form exp jt in 150 mm conc path & insert 138 mm × 12 mm fib strip sld at top wi 12 mm ×. 12 mm bit compd.	Form expansion joint in 150 mm concrete path and insert 138 mm × 12 mm fibre strip sealed at top with 12 mm × 12 mm bituminous compound.

D
Brickwork and Blockwork

225 mm brwk c.o.p. to 1.50 m rad in c.l.m.	225 mm brickwork circular on plan to 1.50 m radius in cement lime mortar.
E.o. comm brwk for facg bks & ptg in tinted c.m. wi a nt wthd jt.	Extra over common brickwork for facing bricks and pointing in tinted cement mortar with a neat weathered joint.
100 mm c. blwk in c.m. blt o/hd.	100 mm concrete blockwork in cement mortar built overhand.

B.i. end of 1. met brr to blwk & m.g. f.f. o.s.	Build in end of large metal bearer to blockwork and make good fair face one side.
C. & p. end of r.s.j. to xstg brwk & m.g.	Cut and pin end of rolled steel joist to existing brickwork and make good.
W. & p. 150 mm blwk up to xstg susp conc clg.	Wedge and pin 150 mm blockwork up to existing suspended concrete ceiling.
C.t. & b. 225 mm brwk to xstg st wallg.	Cut, tooth and bond 225 mm brickwork to existing stone walling.
B.b. 100 mm c. blwk to xstg h.p. wallg.	Block bond 100 mm concrete blockwork to existing hollow pot walling.
Hole thro 450 mm brk ret wall for l.p.	Hole through 450 mm brick retaining wall for large pipe.
225 mm × 150 mm conc a.b. & b.i. to o.s. of h.w.	225 mm × 150 mm concrete air brick and build in to outer skin of hollow wall.
L. & p. 225 mm brwk for raisg.	Level and prepare 225 mm brickwork for raising.
Bit felt d.p.c. to 100 mm wall.	Bituminous felt damp proof course to 100 mm wall.

E
Roofing

Pl conc roofg tiles fxd to s/wd battens at 100 mm c/c wi cu nls.	Plain concrete roofing tiles fixed to softwood battens at 100 mm centres with copper nails.
R.c. on tiles at abutmts.	Raking cutting on tiles at abutments.
C. & f. tiles ard x. 1. p.	Cut and fit tiles around extra large pipe.
H.r. conc ridge tiles bdd & ptd in c.m.	Half round concrete ridge tiles bedded and pointed in cement mortar.
F.e. to ridge tile	Fair end to ridge tile.
1.8 mm lead flashg 225 mm gth dressed over tiles & wi t.e. t. i. brk jt incl r.o. jt & l.w. at 150 mm c/c.	1.8 mm lead flashing 225 mm girth dressed over tiles and with top edge turned into brick joint including raking out joint and lead wedging at 150 mm centres.

F
Carpentry, Joinery and Ironmongery

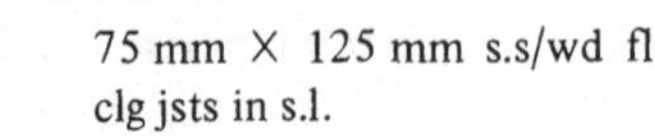

Tan s.s/wd ptn frmg.	Tanalised sawn softwood partition framing.
25 mm × 50 mm s. s/wd brktg shot fired to st col.	25 mm × 50 mm softwood bracketing shot fired to steel column.
75 mm × 125 mm s.s/wd flr & clg jsts in s.l.	75 mm × 125 mm sawn softwood floor and ceiling joists in short lengths.
25 mm s.s/wd roof bdg in n.w. wi t. & g. jts.	25 mm sawn softwood roof boarding in narrow widths with tongued and grooved joints.

50 mm s/wd l. & b.dr wi 25 mm V-jtg bdg.	50 mm softwood ledged and braced door with 25 mm V-jointed boarding.
75 mm × 100 mm s/wd 2ce reb & 1ce grvd dr fr p. & s. to stwk.	75 mm × 100 mm softwood twice rebated and once grooved door frame plugged and screwed to stonework.
25 mm × 125 mm s/wd ling t. at L s.	25 mm × 125 mm softwood lining tongued at angles.
38 mm × 63 mm s/wd splyd & rdd arch.	38 mm × 63 mm softwood splayed and rounded architrave.
25 mm × 100 mm s/wd sktg on & incl sm s.s/wd grds p. to w.	25 mm × 100 mm softwood skirting on and including small sawn softwood grounds plugged to wall.
25 mm × 50 mm h/wd glg bd fxd to frmg wi br c. & s.	25 mm × 50 mm hardwood glazing bead fixed to framing with brass cups and screws.
100 mm p.s. butts scwd to b/bd.	100 mm pressed steel butts screwed to blockboard.
150 mm c.p.br b/bolt scwd to chbd wi skt l.i. conc.	150 mm chromium plated brass barrel bolt screwed to chipboard with socket let into concrete.
D.f.s. & f.o.m.l. & furn to int skel cored fl dr.	Draw from store and fix only mortice lock and furniture to internal skeleton cored flush door.
Cyl n.l. incl perf thro 50 mm th ext h/wd dr.	Cylinder night latch including perforation through 50 mm thick external hardwood door.
S.a.a. csmt stay fxd to met wi s-t screws.	Silver anodised aluminium casement stay fixed to metal with self-tapping screws.

G

Steelwork and Metalwork

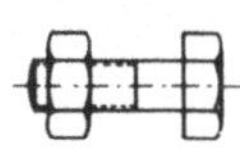

Struct stlwk c/o 150 mm × 150 mm univ cols, 150 mm × 75 mm r.s. chs. & 100 mm × 75 mm m.s. L ties, c/w clts, b.plates & bl lolts wi hex hd, n. &w.	Structural steelwork consisting of 150 mm × 150 mm universal columns, 150 mm × 75 mm rolled steel channels and 100 mm × 75 mm mild steel angle ties complete with cleats, base plates and black bolts with hexagonal head, nut and washer.
Compd bm m.u. of 2 No. r.s. js, 2 No r.s. conn plates & sq. hdd bolts, n. & w. incl all nec drillg.	Compound beam made up of 2 No. rolled steel joists, 2 No. rolled steel connecting plates and square headed bolts, nuts and washers, including all necessary drilling.
75 mm × 50 mm galv m.s. T-suppt bolted to wall (bolts & holes m.s.)	75 mm × 50 mm galvanised mild steel tee-support bolted to wall (bolts and holes measured separately).
38 mm × 6 mm wrot iron fl h/rl fxd on rk.	38 mm × 6 mm wrought iron flat handrail fixed on rake.
P-m alum brkt compr 300 mm len 25 mm × 4 mm strip 3ce bt, 1ce ntchd & 4ce csk drilled.	Purpose-made aluminium bracket comprising a 300 mm length of 25 mm × 4 mm strip 3 times bent, once notched and 4 times countersunk drilled.

12 mm diam h.d. bolt & set in pos.	12 mm diameter holding down bolt and set in position.
Sher met wdw 1,000 mm × 1,500 mm o/all p. & s. to hd & jmbs of opg.	Sherardised metal window 1,000 mm × 1,500 mm overall plugged and screwed to head and jambs of opening.

H

Plumbing

100 mm diam pvc r.w.p. c/w brkts p. & s. to rec st wallg.	100 mm diameter polyvinylchloride rainwater pipe complete with brackets plugged and screwed to reconstructed stone walling.
150 mm h.r. p.s. e.g. c/w brkts scrwd to s/wd fascia bd.	150 mm half round pressed steel eaves gutter complete with brackets screwed to softwood fascia board.
22 mm o.d. cu tubg in c.w.s.	22 mm outside diameter copper tubing in cold water service.
22 mm ditto in h.w.s. fxd in p. csg.	22 mm outside diameter copper tubing in hot water service fixed in pipe casing.
28 mm o.d. cu o/flow.	28 mm outside diameter copper overflow.
150 mm i.d. c.i. s. & v. p. wi ears c.o. p. & s. to conc stan csg.	150 mm internal diameter cast iron soil and ventilation pipe with ears cast on plugged and screwed to concrete stanchion casing.
22 mm br h.p. s.d. s.c. wi compr jts b/e to pe tubg.	22 mm brass high pressure screw down stop cock with compression joints both ends to polythene tubing.
L.l.w.c. suite c/o wh vit china pan wi pl seat, 9 lit cap cist wi 15 mm diam l.p. b/v & 22 mm diam sh fl p., incl fxg to wall wi spec brkts & jtg to dr.	Low level w.c. suite consisting of a white vitreous china pan with plastic seat, 9 litre capacity cistern with 15 mm diameter low pressure ball valve and 22 mm diameter short flush pipe, including fixing to wall with special brackets and jointing to drain.

J

Plastering etc.

R. & s. in plas to wall rvls n.e. 100 mm wd.	Render and set in plaster to wall reveals not exceeding 100 mm wide.
R.f. & s. in plas to clgs exc 3.5m hi.	Render float and set in plaster to ceilings exceeding 3.5 m high.
12 mm p/bd fxd to s. & s. bm. & skim ct of plas.	12 mm plasterboard fixed to sides and soffites of beam and skimming coat of plaster.
Lab f.e. & a. on plas.	Labour fair edge and arris on plaster.
50 mm grano pavg (1 : 1 : 2) ld to falls on o/s conc.	50 mm granolithic paving (1 : 1 : 2) laid to falls on oversite concrete.
150 mm × 150 mm × 16 mm th q.t. pavg bdd, jtd & ptd in c.m.	150 mm × 150 mm × 16 mm thick quarry tile paving bedded, jointed and pointed in cement mortar.

F. sq c.on q.t. agst met fr.	Fair square cutting on quarry tiles against metal frame.
100 mm × 100 mm × 6 mm w.g. w.t. stk to plas wall wi apprd adhes.	100 mm × 100 mm × 6 mm white glazed wall tiles stuck to plastered wall with approved adhesive.
R.e. on tiles.	Rounded edge on tiles.

K
Glazing

4 mm o.q.c.s. gl & glg to met wi putty in pns exc 1.00 m^2.	4 mm ordinary quality clear sheet glass and glazing to metal with putty in panes exceeding 1.00 m^2.
Apprvd obs patt gl & glg to wd wi bds (m.s.) in pns n.e. 0.10 m^2.	Approved obscured pattern glass and glazing to wood with beads (measured separately) in panes not exceeding 0,10 m^2.
6 mm G. w.c. gl & glg to pat glg brs of roof lights.	6 mm Georgian wired cast glass and glazing to patent glazing bars of roof lights.
6 mm pol pl gl & glg to h/wd reb wi bds (m.s.).	6 mm polished plate glass and glazing to hardwood rebate with beads (measured separately).
W.l. strip & embddg edge of gl therein.	Washleather strip and embedding edge of glass therein.

L
Painting and Decorating

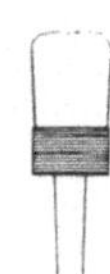

Prep & 2 cts emul on ct rend wall surfs int.	Prepare and apply 2 coats of emulsion paint on cement rendered wall surfaces internally.
Prep. sl & (2) on f.f. brwk ext.	Prepare, seal and apply 2 coats of oil paint on fair faced brickwork externally.
W.d. old wdwk int & 1 u/ct & 1 gl f/ct oil pt on g.s.	Wash down old woodwork internally and apply 1 undercoat and 1 gloss finishing coat of oil paint on general surfaces.
K.p.s. & (3) wd surfs ext 100 mm–200 mm gth.	Knot, prime, stop and apply 3 coats of oil paint on wood surfaces externally exceeding 100 mm and not exceeding 200 mm girth.
St & wax pol h/wd surfs int 200 mm–300 mm gth.	Stain and wax polish hardwood surfaces internally exceeding 200 mm and not exceeding 300 mm girth.
T.u.pr ct & (2) h.r. int on htg surfs rads.	Touch up primer coat and apply 2 coats heat resisting oil paint internally on heating surfaces of radiators.
H.o. vinyl w/paper (prov by client) to plas walls wi h.d. adhes.	Hang only vinyl wallpaper (provided by client) to plastered walls with heavy duty adhesive.

Vocabulary	Wörterverzeichnis	Glossaire
construction term	Ausdruck aus dem Baufach	terme de construction
surveyor, quantity surveyor	Fachingenieur, der Bauten finanziell abwickelt	ingénieur spécialisé dans les décomptes
to take off quantities	Massen herausziehen	enlever des quantités
bills of quantities	Leistungsverzeichnis	tableau des quantités
site measurement	Baustellenaufmaß	métrage du chantier
final account	Endabrechnung	décompte final
damp proof course (d.p.c.)	waagerechte Sperrschicht	couche d'étanchéité horizontale
ground floor (g/f)	Erdgeschoß	rez-de-chaussée
to mark up, (to write the names, sizes, etc. of the materials on the drawing)	Zeichnungen mit näheren Materialangaben versehen	marquer
trade	Handwerk	artisanat, arts

A.

demolition	Abbruch, Zertrümmerung	démolition
to break up	abbrechen, zerkleinern	casser
concrete foundation	Betonfundament	fondation en béton
to remove	wegschaffen, beseitigen	enlever
to break out	herausbrechen	découper, sortir avec violence
cast iron drain	gußeisernes Abflußrohr	tuyau de fuite en fonte
to take up	aufnehmen, aufreißen	faire le relevé
stone paving	Steinpflaster	pavé en pierre
to set aside	beiseite setzen	mettre de côté
to pull down	niederreißen, einreißen	abattre
brickwork	Ziegelmauerwerk	maçonnerie en briques
to cut out	ausbrechen	découper
ventilator	Lüftungsöffnung	ventilateur
hollow wall	zweischalige Wand, Hohlwand	paroi creuse
to make good	sauber nacharbeiten	curer
cement rendering	Zementputz	enduit de ciment
to take off	abnehmen, abmontieren	démonter
rim lock	(Aufsatz-)Kastenschloß	serrure à coffre
furniture	Drückergarnitur, Türschild mit Drücker, Beschlag	garniture
to return to store	ins Lager zurückbringen	retourner au magasin
to hack off	abschlagen	abattre
plaster	Innenputz	enduit intérieur
blockwork	großformatiges (Beton- oder Naturstein-)Mauerwerk	maçonnerie en blocs
to take down	abreißen, abtragen	démolir
plasterboard, gypsum plasterboard	Gipskartonplatte	panneau en carton-plâtre
plasterboarded	mit Gipskartonplatten versehen	muni de panneaux en carton-plâtre

B.

to excavate	(Boden) ausheben, ausschachten	creuser
over site	im Bereich der Baustelle	dans la zone du chantier
to deposit	absetzen, abladen	déposer
spoil heap	Haufen überschüssigen Aushubs	décharge
to reduce levels	Höhen abtragen	abaisser les niveaux
trench	Graben	tranchée

extra over	Zulage	supplément
to fill and ram	verfüllen und abstampfen	remblayer et compacter
to level	einebnen, planieren	niveler
bottom	Sohle	fond
hardcore	Schüttmaterial	blocage
to spread and level	verteilen und einebnen	répartir et niveler
to plank and strut	einbohlen und aussteifen	coffrer et raidir
basement	Kellergeschoß	sous-sol
to roll and consolidate	abwalzen und verdichten	cylindrer et compacter
blinding with fine material	Oberfläche verschließen mit feinem Füllmaterial	sceller la surface avec du matériau fin
oversite concrete	Unterbeton	béton de semelle

C.

reinforced concrete	Stahlbeton	béton armé
ground beam	Fundamentbalken	poutre de fondation
sectional area	Querschnittsfläche	surface de la coupe transversale
plain concrete	unbewehrter Beton	béton simple
ground floor slab	Sohlplatte	radier
cement and sand bed	Betonestrich, Zementestrich	chape en ciment
trowel	Kelle	truelle
to trowel concrete	Beton mit der Kelle glätten	lisser du béton à la truelle
paving	Pflaster, Wegoberfläche	revêtement d'une voie
smooth as paving	gut geglättet	bien lisse
precast concrete lintol	Fertigbetonsturz	linteau préfabriqué en béton
finished fair on exposed surfaces	mit sauber gearbeiteten Sichtflächen	avec surfaces apparentes de bonne qualité
mild steel	Flußstahl	acier doux
reinforcing rod	Betonstahlstab	barre pour béton armé
wrought formwork	gehobelte Schalung	coffrage raboté
soffite	Unterfläche	sous-face
concrete beam	Betonbalken	poutre en béton
fillet	Leiste	baguette
to plant on, (to fix with nails or pins)	anbringen, mit Nägeln befestigen	fixer avec des clous
throating	Wassernase	gouttière pour eau de condensation
window head	Fensterhaupt	linteau
(form) mortice	Zapfenloch (herstellen)	mortaiser
landing	Treppenabsatz	palier
baluster	Treppenstab, Tralje	balustre
to grout in	einmörteln	garnir de mortier
mortar	Mörtel	mortier
expansion joint	Dehnungsfuge	joint de dilatation
path	Fußweg	sentier
to insert	einlegen	insérer
fibre strip	Streifen aus Fasermaterial	bande en fibres
sealed	versiegelt, vergossen	scellé
bituminous compound	Bitumenmasse	mélange bitumineux

D.

brickwork	Ziegelmauerwerk	maçonnerie en briques
circular	kreisförmig	circulaire
on plan	im Grundriß	dans le plan horizontal
cement lime mortar	Kalkzementmörtel	mortier de chaux et ciment

common brickwork	Mauerwerk aus Normalziegeln	maçonnerie en briques normales
facing brick	Vormauerziegel	brique de revêtement
to point	ausfugen	jointoyer
tinted	gefärbt	teinté
weathered joint	unterschnittene Fuge	joint coupé
to build overhand	über Hand mauern	maçonner à contre-main
metal bearer	Stahlträger	poutre en acier
to make good fair face	Sichtfläche einwandfrei herstellen, beiarbeiten	bien surfacer
to cut and pin	abschneiden, und festsetzen	couper et fixer
rolled steel joist	I (Walz-) Träger	poutre en acier laminé en I
to make good	beiarbeiten	retoucher
to wedge and pin	verkeilen und festsetzen	coincer et fixer
suspended concrete ceiling	freigespannte Betondecke	plancher en béton à portée libre
to cut, tooth and bond	ausstemmen und verzahnt einbinden	creuser au ciseau et endenter
to block bond	im Blockverband verbinden	relier en blocs
hollow pot walling	Hohlblockstein-Wand	mur en blocs creux
retaining wall	Stützmauer	mur de soutènement
large pipe	weites Rohr	tube à grand diamètre
concrete air brick	Gasbetonstein, Schaumbetonstein	aggloméré de ventilation
outer skin	Außenschale	paroi éxterieure du mur
hollow wall	zweischalige Wand	paroi creuse
to raise	aufmauern, erhöhen	réhausser
bituminous felt	Bitumenpappe	carton bitumé
damp proof course (d.p.c.)	Sperrschicht	couche imperméable

E.

roofing tile	Dachziegel	tuile
softwood batten	Dachlatte aus Weichholz	latte en bois tendre
at centres	mit Abstand, Mittenabstand	d'axe en axe
copper nail	Kupfernagel	clou en cuivre
to cut raking, to cut on rake	schräg abtrennen	couper en biais
abutment	Trennlinie (Überschneidung) von z. B. Dach- mit Wandfläche	aboutement
to cut and fit	schneiden und einpassen	couper et ajuster
ridge tile	Firstziegel	tuile faîtière
(form) fair end	sauberen Abschluß herstellen	bien finir
lead flashing	Dichtungsblech aus Blei, Bleieinfassung	feuille d'étanchéité en plomb, bordure en plomb
girth	Umfang, gesamtes Außenmaß	périmètre
to dress	bekleiden, bedecken	habiller
brick joint	Mauerfuge	joint de maçonnerie
top edge	obere Kante	arête supérieure
to rake out	auskratzen	gratter
lead wedging	mit Bleiverkeilung	calage avec du plomp

F.

carpentry	Zimmerarbeiten	charpenterie
joinery	Tischlerarbeiten	menuiserie
ironmongery	Schlosserarbeiten	serrurerie
to tanalise	mit Holzschutzmittel behandeln	traiter avec un produit de protection du bois
sawn softwood	Schnitt(weich)holz	bois de sciage tendre
partition framing	Zwischenwand-Rahmenkonstruktion, Skelett	ossature de cloisonnement

softwood bracketing shot fired to steel column	Montageleisten aus Weichholz, an Stahlstütze angeschossen	baguettes de montage en bois tendre rattachées aux poteaux en acier
floor and ceiling joists	Fußboden- und Deckenbalken	poutres de plancher
roof boarding	Dachverschalung	coffrage de toit
narrow width	schmal	étroit
tongued and grooved joint	Nut- und Federverbindung, Falzspundung	bouvetage
ledge	Querholz	traverse
brace	Strebe	buton
ledged and braced door	(abgestrebte) Brettertür	porte à membrure de renforcement
V-jointed boarding	Verbretterung in Gratspundung	planchéiage à joints en V
rebated	gefalzt	bouveté
grooved	genutet	rainuré et languetté
door frame	Türrahmen	huisserie
plugged and screwed to stonework	m.H. von Dübeln an Steinmauerwerk angeschraubt	vissé dans la maçonnerie à l'aide de chevilles
(door) lining	(Tür-)Bekleidung	panneaux de porte
tongued at angles	auf Gehrung und mit Federverbindung	languetté et biseauté
splayed and rounded	(eine Kante) abgeschrägt und (die andere Kante) abgerundet	biseauté d'un côte et arrondi de l'autre
architrave	Deckleiste	architrave
skirting	Fußleiste, Scheuerleiste	plinthe
softwood ground	Holzdübelstück	cheville en bois
to plug	dübeln	cheviller
glazing bead	Falzleiste	latte à vitrage
brass cup	Unterlegrosette aus Messing	rondelle en laiton
pressed steel butt	Scharnier aus Stahlblech	charnière en acier
blockboard	Tischlerplatte	panneau en latte
chromium plated	verchromt	chromé
barrel bolt	Grendel-Riegel	verrou
chipboard	Spanplatte	panneau en copeaux
socket	Steinbuchse, Einlaßhülse	manchon
to draw from store	"bauseitig geliefert", vom Lager holen	fourni par l'entrepreneur
mortice lock	Einsteckschloß	serrure encastrée
furniture	Beschlag, Türschilder und Drücker	ferrures, plaque de porte et ouvre-porte
skeleton cored flush door	abgesperrte Tür	porte lisse à ossature
cylinder night latch	Zylinder-Kasten-Fallenschloß	serrure dormante cylindrique
perforation	Bohrung	perçage
external hardwood door	Außentür aus Hartholz	porte extérieure en bois dur
silver anodised aluminium	silberfarbig eloxiert	aluminium anodisé de teinte argentée
casement stay	Fensterfeststeller, Kieler Stellstange	blocage de fenêtre
self-tapping screw	selbstschneidende Schraube, Blechtreibschraube	vis autocoupeuse

G.

steelwork	Stahlarbeiten	travail de l'acier
metalwork	(sonstige) Metallarbeiten	travail des métaux
structural steelwork	Stahlbauarbeiten	travaux de construction en acier

universal column	Breitflanschträger	poutre à semelles larges
rolled steel channel	U-Profil, U-Eisen	profil en U
mild steel angle tie	L-Profil-Zugstab	acier doux profilé en L
cleat	Anschlußwinkel, Auflagerwinkel	cornière d'assemblage
base plate	Fußplatte	plaque de base
black bolt	rohe Schraube	vis non travaillée
hexagonal head	Sechskantkopf	six pans
nut	Mutter	écrou
washer	Unterlegscheibe	rondelle
compound beam	zusammengesetzter Träger	poutre assemblée
rolled steel joist	I-Träger (warm gewalzt)	poutre en I laminée à chaud
connecting plate	Bindeblech	tôle de liaison
square headed bolt	Schraube mit Vierkant-Kopf	vis à tête à 4 pans
to drill	bohren	forer
galvanised	verzinkt	galvanisé
tee-support	T-Profil-Auflager	support en T
bolted to wall	an die Wand geschraubt	vissé au mur
wrought iron	Schmiedeeisen	fer forgé
handrail	Handlauf	main courante
fixed on rake	schräg angebracht	fixé en biais
purpose-made	Sonderanfertigung, spezialangefertigt	fait à la demande
bracket	Befestigungsklammer oder -Schuh, (Kleinteil für Verbindungszwecke)	console de fixation, collier, (petite pièce de liaison)
aluminium strip	Alu-Band	bande en aluminium
to bend	biegen	plier
to notch	ausklinken	entailler
to countersink	versenken	encastrer, foncer, fraiser, noyer
holding-down bolt	Steinschraube	boulon d'ancrage
to sherardise	im Spezialverfahren verzinken	sherardiser
jambs	vertikale Leibungsflächen	jambage

H.

plumbing	Klempner- und Rohrlegearbeiten	plomberie
rainwater pipe	Regenrohr	gouttière
(pipe) bracket	Rohrschelle	collet de gouttière
reconstructed stone	Betonstein mit imitierter Natursteinoberfläche	pierre reconstituée
pressed steel	Stahlblech, kalt oder warm verformt	acier laminé
eaves gutter	Regenrinne	caniveau pour eaux de pluie
fascia board	Traufbrett	chéneau
outside diameter	Außendurchmesser	diamètre extérieur
copper tubing	Kupferrohrleitungen	tuyaux en cuivre
cold water service	Trinkwasser(versorgungs)leitung	conduite d'eau potable
pipe casing	Rohrverkleidung	gaine (habillage) de tuyau
overflow	Überlauf	trop-plein
internal diameter	lichter Durchmesser	diamètre intérieur
soil and ventilation pipe	Abflußrohr, Abfall/Lüftungsrohr	tuyau de fuite et de ventilation
with ears cast on	mit angegossenen (Befestigungs-) Laschen	avec pattes venues de fonderie
concrete stanchion casing	Stahlstützen-Betonummantelung	revêtement en béton à poteaux en acier

screw down stop cock	(Durchgangs-)Ventil mit Spindel	soupape droite à tige filetée
compression joint	Quetschkonusverbinder	joint à compression
polythene tubing	Polyäthylen-Rohrleitungen	tuyaux en polythéne
w.c.suite	Abortbecken-Garnitur	équipement de WC
white vitreous china pan	Abortbecken aus weißem Porzellan	cuvette de WC en porcelaine
plastic seat	Ringsitz (Brille) aus Kunststoff	siège de WC en plastique
cistern	Spülkasten	réservoir de chasse d'eau
low pressure ball valve	Niederdruck-Schwimmerventil	soupape sphérique à basse pression
flush pipe	Spülrohr	tuyau de chasse
to joint to drain	an Abflußrohr anschließen	raccorder au tuyau de descente
J.		
to plaster	putzen (Innenputz, Kalkputz)	enduire intérieur
to render	putzen (Außenputz, Zementputz)	enduire extérieur
render and set in plaster	Unterputz Mörtelgruppe III (II) Oberspurzschicht Mö.Gr.I, zweilagiger Putz	sous-enduit et enduit
reveal	sichtbare Leibungsfläche	intrados
float	Reibebrett	bouclier
render, float and set in plaster	Spritzbewurf Mö.Gr.III sowie zwei Lagen Mö.Gr.I, glatter Putz auf zwei Lagen Unterputz	sous-enduit, enduit et dressage
plasterboard	Gipskartonplatte	panneau carton-plâtre
soffite of beam	Balkenunterseite	sous-face de poutre
skimming coat of plaster	dünne Putzschicht	enduit mince
labour fair edge and arris	sorgfältig Kanten und Ecken beiarbeiten	angles rentrants et arêtes réalisés soigneusement
granolithic paving	Hartbetondeckschicht	revêtement en béton dur
to lay to falls	in Neigung herstellen	exécuter avec inclinaison
oversite concrete	Unterbeton	béton de semelle
quarry tile	Tonfliese, Klinkerplatte	carreau en terre cuite
bedded in cement mortar	in Zementmörtel verlegt	posé sur lit de mortier de ciment
jointed and pointed	verfugt und verstrichen	jointoyé et taloché
fair square cutting	sauberes, gerades Beschneiden	coupe droite et propre
glazed wall tile	glasierte Wandfliese	carreau mural vitrifié
to stick	kleben	coller
approved adhesive	anerkanntes Klebemittel	produit de collage reconnu
rounded edge	abgerundete Kante	angle arrondi
K.		
ordinary quality clear sheet glass	Fensterglas 1.Sorte	verre à vitres ordinaire de 1ere qualité
to glaze to metal	in Metallrahmen einglasen	vitrer sur cadre métallique
putty	Kitt	mastic
pane	Scheibe	vitre
obscured pattern glass	Ornamentglas, Kathedralglas weiß	verre cathédrale opalin, verre ornemental
bead	Falzleiste	latte à vitrage
Georgian wired cast glass	dickes Drahtglas	verre armé épais
patent glazing	kittlose Verglasung	vitrage sans mastic
metal bar	Fenstersprosse aus Metall	petit bois de fenêtre en métal

roof light	Oberlicht	vasistas
polished plate glass	Spiegelglas	verre à glace
rebate	Falz, Anschlag	feuillure
L.		
painting	Malerarbeiten	travaux de peinture
decorating	Anstreicher- und Tapezierarbeiten	travaux de peinture et de tapisserie
coat of emulsion paint	Emulsionsfarbanstrich	couche de peinture émulsionnée
to seal	versiegeln	sceller
to apply 2 coats of oil paint	2 mal mit Ölfarbe streichen	peindre à deux couches de peinture à l'huile
faced brickwork	Vorziegelmauerwerk	maçonnerie de parement en briques
to wash down	abwaschen	laver
undercoat	Grundanstrich	couche de fond
gloss finishing coat of oil paint	Schlußanstrich mit Standölfarbe	couche de finissage à la peinture à l'huile
to knot	absperren harzhaltiger Äste	sceller des noeuds résineux
to prime	grundieren	donner la couche d'impression,
to stop	verkitten, spachteln von Fehlstellen	spatuler
to stain	beizen	mordancer
to touch up	aufbessern, auffrischen	retoucher
heat resisting paint	Heizkörperfarbe	peinture à radiateurs
radiator	Heizkörper	radiateur
to hang	anbringen, ankleben	pendre
vinyl wallpaper	Tapete mit waschfester Oberfläche	tapisserie murale en plastique

COMMON BRITISH ABBREVIATIONS FOR PLASTICS USED IN THE CONSTRUCTION INDUSTRY

Abbreviation	Full Name	Some Typical Applications
a.b.s.	acrylonitile butadiene styrene.	ducting, waste and high-strength pipes and wall panelling.
g.r.p.	glass fibre reinforced polyester resin.	sanitary fittings, windows, roofing sheets and wall panelling.
pe.	polyethylene or polythene.	damp proof membranes, cold water pipes and storage cisterns.
pm.ma.	polymethyl methacrylate or acrylic.	rooflights, signs, sanitary fittings, wall panelling and lighting fittings.
pp.	polypropylene.	waste traps, w.c.seats and drain pipe couplings.
pr.	polyester.	rooflights, translucent roofing sheets and wall panelling, electrical appliances and hard-wearing floor finishings.
ps.	polystyrene.	thermal insulation sheets, pipe lagging, wall and ceiling tiles, storage cisterns, ventilators and gratings.
ptfe.	polytetrafluorethylene.	electrical insulators and cable sheating, linings for tanks etc. and pipe thread sealing tape.
pu.	polyurethane.	seamless floor finishing, lacquers and paints, electric cable sheathing and protective site clothing.
pvc.	polyvinylchloride.	rainwater and soil systems, floor tiles and sheet coverings, ducting, translucent roofing sheets and wall panelling and electric cable sheathing.

Vokabulary	Wörterverzeichnis	Glossaire
plastics	Kunststoff	plastique
construction industry	Baugewerbe, Bauindustrie	industrie du bâtiment
acrylonitile butadiene styrene (a.b.s.)	Acrylnitril-Butadien-Styrol-Copolymere (ABS)	acrylenitrile-butadiène-styrène-copolymère
ducting, duct	abgedeckter Kanal	gaine
waste pipe	Abwasserrohr	tuyau d'eaux usées
wall panelling	(Leicht-)Wandelement	revêtement mural en panneaux
glass fibre reinforced polyester resin (g.r.p.)	glasfaserverstärktes Polyesterharz (GUP)	polyester renforcé à la fibre de verre (PRE)
sanitary fitting	sanitäres Formstück	petite pièce d'équipement sanitaire
roofing sheet	Dachdeckungstafel, ~platte	plaques de couverture
polyethylene, polythene (pe.)	Polyäthylen (PE)	polythène
damp proof membrane	Hautabdichtung, Sperrschicht	voile d'étanchéité
storage cistern	Speicher(behälter)	réservoir de stockage
polymethyl methacrylate, acrylic (pm.ma.)	Polymethacrylsäuremethylester, Acrylglas (PMMA)	verre acrylique (PMMA)

rooflight	Oberlicht	vasistas
sign, signboard	Schild	panneau indicateur
lighting fitting	Beleuchtungskörper-Armatur, Lampenfassung	luminaire
polypropylene (pp.)	Polypropylen (PP)	polypropylène
waste trap	Geruchverschluß	siphon coupe-air
w.c.seat	Abortsitz	siège de WC
drain pipe coupling	Abwasserrohrverbinder	embranchement de tuyaux d'eaux usées
polyester (pr.)	Polyester (PETP, PC)	polyester
translucent roofing sheet	lichtdurchlässige Dachdeckungstafel	plaque de couverture translucide
electrical appliance	elektrisches Gerät	appareil électrique
hard-wearing floor finishing	Industrie-Bodenbelag, hochbeanspruchbare Fußbodenbeschichtung	revêtement de sol à haute résistance
polystyrene (ps.)	Polystyrol (PS)	polystyrène
thermal insulation sheet	Dämmplatte	couche isothermique
pipe lagging	Rohrummantelung	isolation des tuyaux
wall tile	Wandfliese	carreau mural
ventilator, vent	Ent- und Belüftungsöffnung	ouverture de ventilation
grating	Abdeckrost	grille de fermeture
polytetrafluorethylene (ptfe.)	Polytetrafluoräthylen (PTFE)	polytétrafluoréthylène
electrical insulator	elektrischer Isolator	isolateur électrique
cable sheathing	Kabelmantel	enveloppe de câble
lining for tank	Beschichtung von Behältern	stratification de réservoirs
pipe thread sealing tape	Rohrgewindedichtungsband	ruban d'étanchéisation de filetage de tuyau
polyurethane (pu.)	Polyurethan (PUR)	polyuréthane
seamless floor finishing	fugenloser Fußbodenbelag	revêtement de sol sans joints
lacquer	Lack	laque
paint	Farbe	peinture
protective site clothing	Arbeitsschutzbekleidung für Baustelle	vêtement de protection pour chantier
polyvinylchloride (pvc.)	Polyvinylchlorid (PVC)	polyvinylchloride (PVC)
rainwater and soil systems	Regenwasser- und Schmutzwasserleitungssysteme	conduites d'eaux de pluie et d'eaux usées
floor sheet covering	Fußbodenbelag	revêtement de sol